SOYUZ

1967 onwards (all models)

Owner's Workshop Manual

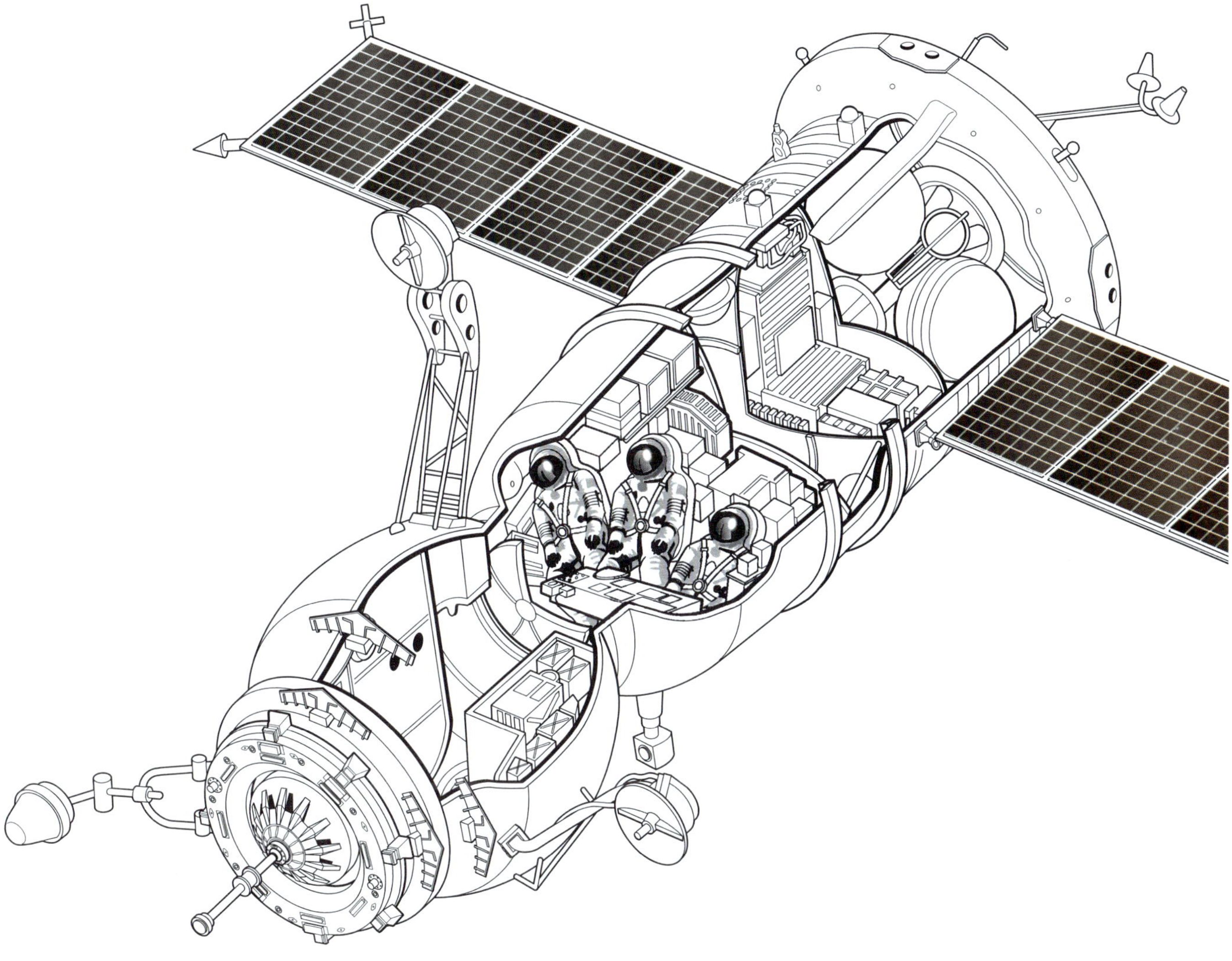

An insight into Russia's flagship spacecraft from Moon missions to the International Space Station

David Baker

Foreword by the first British astronaut, Dr. Helen Sharman OBE

Contents

3	Foreword

4	A world lead

Shaky start 6
To the cosmos! 7
Dream of big things 8

16	Vostok and Voskhod

Man in space 23

46	The 7K-OK Soyuz

Korolev's train 50
A 'go' for the Moon 60
Redirection 64

Feature A The Soyuz spacecraft 68

98	Soyuz 7K-OK and Zond flights 1966–71

Tragedy strikes 101
The Soyuz gunship 103
Recovery 108
Back to space 111
A change of direction 118

Feature B Rendezvous and docking 120

128	Soyuz 7K-T ferry flights 1973–81

Salyut 130
Second-generation K-T series flights 133

Feature C A handshake in space 138

150	Soyuz T 1979–86

Feature D Progress 7K-TG, 7K-TGM, 7K-TGM1 155

160	Soyuz TM 1986–2002

166	Soyuz TMA/TMA-M 2002–??

Soyuz TMA 166
Soyuz TMA-M 168

170	Abbreviations

170	Index

Foreword

Making one half-turn of the adjuster on the carburettor of my first motorbike, I studied the well-thumbed pages of the Haynes Manual lying on the ground beside me.

In the days when engines were relatively simple and a novice like me could tackle such things as the fuel mixture without the aid of advanced electronic testing equipment, my Haynes Manual kept me on the road. Seven years later, training to be an astronaut in Star City near Moscow, I would long for something as detailed and complete for the Soyuz-TM12 as I had back then for my Honda CD-175.

There are history books and technical papers about spacecraft components and systems, but never before have I seen anything that sets the political context and human endeavour alongside the technology required for the advancement of space flight. And context is important, even for scientists and engineers. I remember asking one of my trainers why the Globus positioning system was connected to other systems in the way it is, to be told that in order to maintain a comfortable element of safety, new systems tend to be added to those that have already proved themselves. At least that way, you can rely on at least so much of the spacecraft working. Anyway, I was told, it had worked well for Gagarin and would I really want to be the first to test out something totally new? I always thought of Gagarin when I used Globus after that.

Nowadays, making a Soyuz spacecraft has become quite routine, almost a production line operation. David Baker relates how lessons learned from early failures – for which some paid with their lives – have produced a hugely reliable spacecraft and launching system. Now, the Chinese look set to base lunar landings on similar technology, something that must come as bittersweet irony to the Soviet engineers and cosmonauts who were working towards lunar missions in the 1960s. (The Soviets were the first in space and the first to carry out a spacewalk. First on the Moon would have been the crowning glory.)

You could describe Soyuz as a workhorse. I remember it affectionately as my home for a couple of days and my safe retreat in case of an emergency on board Mir. I hope that history will accord Soyuz the respect it deserves: as David says, it is flexible, adaptable and has sustained several generations of development. For this and its reliable, constant support of a range of activities that are themselves internationally significant, Soyuz deserves regal status. Of all spacecraft, Soyuz is Queen.

Helen Sharman
January 2014

Chapter One

A world lead

When Russia launched the first artificial Earth-orbiting satellite on 4 October 1957, it changed the world and heralded the dawn of the Space Age. Known to its engineers as PS-1, the first Russian satellite was officially named Sputnik 1.

OPPOSITE Launched on 4 October 1957, Sputnik 1 represented a major achievement for the Soviet Union and propelled the world into the Space Age in which the next great technological race would be run. *(David Baker)*

RIGHT Sergei Korolev (1907–66) was the brain behind the R-7 ballistic missile and the early Soviet space programme, the architect of Russia's satellite programmes and the instigator of human space flight in the USSR. *(David Baker)*

Within a month a second satellite, PS-2 (Sputnik 2), had been launched, carrying a dog named Laika. It had been hurriedly prepared on the enthusiastic insistence of Soviet Premier Nikita Khrushchev, responding to rapturous applause from the Russian citizenry. There was no means of bringing Laika back to Earth and within four days she died of heat exhaustion.

Sputnik 1 had been a spherical object weighing 184lb (83.6kg) with a diameter of 23in (58cm), but its successor weighed 1,120lb (508.3kg). PS-1 and PS-2 were interim projects rushed into space to achieve a coup over the Americans.

The decision to build a satellite had been made in 1955 when the US announced that it planned to launch a satellite for the International Geophysical Year of 1957–58. Not to be outdone – and with the reluctant agreement of Soviet Premier Nikita Khrushchev – the Russians pledged to launch their own satellite, spurred on by a gifted rocket designer, Sergei Korolev. So began the simple satellites that opened a new age for human exploration of the cosmos. It was Korolev too who would be the brains, and the push, behind Russia's first man in space, and eventually the spacecraft that still flies today.

Shaky start

Korolev was born in 1907 in the Ukrainian town of Zhytomir. Pursuing an early interest in aeronautical engineering, he became fascinated with the prospect of rocket-propelled aircraft. After he moved with his parents to Moscow he was tutored by none other than the legendary aircraft designer Andrei Tupolev, after which he joined an aircraft design bureau. From 1931 he was involved with the first government-approved rocket development programme through the Group for the Study of Reactive Flight (GIRD).

Over the next several years Korolev led the GIRD in designing and testing several liquid-propellant rockets and publishing his results to a wider audience. From 1933 the military began funding the GIRD, and eventually it merged with a similar organisation known as the Gas Dynamics Laboratory (GDL) in Leningrad to become the Jet Propulsion Research Institute (RNII). Into this group many of the future Russian rocket engineers were brought, including Valentin Glushko, who would become one of the leading engine builders.

Korolev was hard-working, demanding and yet persuasive in his arguments and demeanour, using his strong personality to ensure success through censure and a refusal to accept shoddy work. This brought

RIGHT Early Russian rocketry was inspired by space prophets, science fiction writers and scientists who grasped the significance of probing beyond Earth's tenuous atmosphere to sample the conditions prevalent on other worlds. *(TASS)*

conflicts with colleagues, especially Glushko, who denounced him during the purges of 1938. Both Korolev and Glushko were in turn denounced by Andrei Kostikov, who became head of the RNII. Tortured in Moscow's notorious Lubyanka prison and sentenced to ten years in a gulag, Korolev was partially reprieved when he wrote direct to Stalin, and succeeded in getting a retrial after Nikolai Yezhov, the hated head of the secret police (the NKVD), was replaced with Lavrenti Beria.

Korolev was lucky to escape with his life. Other leading lights in the RNII were executed. It had been enough to be accused of the 'personality cult' and wasting national resources. In the wave of mass killings many leading rocket engineers lost their lives, and the outstanding work they had conducted was set back by a decade. In Germany, meanwhile, the army put leading rocket engineers to work building a long-range ballistic missile, the A-4 (eventually known as the V-2), just as their Russian counterparts, unaware of the German project, picked up the pieces and resumed work.

Korolev was sent to a special prison for intellectuals and academics and when Germany invaded Russia in June 1941 the rocket engineers were sent to work with CKB-29, a design bureau led by Tupolev. A year later he moved again to another prison run by Glushko. Korolev was fully discharged from the attentions of the NKVD in June 1944 and sent to the aviation commission for assignment, where his fortunes turned for the better. In 1945 he received a Badge of Honour for his contribution to the development of rocket motors for assisting heavy aircraft into the air and was commissioned into the Red Army with the rank of colonel.

Immediately after the war the Russians gathered up all the German rocket scientists they could find to exploit their knowledge and experience. The V-2 represented a considerable advantage for the Russians in that it could provide working examples of a long-range missile capable of sending a one-tonne warhead across a range of up to 200 miles (322km). Stalin ordered full-scale development of bigger and better missiles and by the early 1950s the Red Army had a limited supply of rockets with a range of up to 746 miles (1,200km).

Also during the early 1950s, in the final years

ABOVE Soviet Army officers relax during a hunt for German rocket scientists. At right, arms folded, is Boris Chertok, one of the close group of engineers who would mastermind the early Soviet space programme. *(Boris Chertok)*

of Soviet premier Joseph Stalin, Russian nuclear physicists Igor Kurchatov and Andrei Sakharov frequently met with Korolev and his team to marry the missile to the atomic bomb. But none of these early rockets could carry heavy atomic weapons and none had the range to threaten the United States. Unlike the Russians, the Americans had a very large fleet of long-range heavy bombers each capable of carrying one or more nuclear bombs, while the British were developing their V-bombers to carry atomic war to Moscow.

To the cosmos!

It is no exaggeration to say that four events in 1953 opened the way for Soviet leadership in very long-range rockets and space travel: the death of Stalin, which began an entirely new relationship between the scientists and the Kremlin; the decision to design an intercontinental ballistic missile (ICBM) capable of delivering atomic bombs on targets in the United States across the North Pole; the detonation of Russia's first hydrogen bomb; and the commitment by Korolev to plan the first Soviet steps into space and to persuade the new leadership to do it for science and the prestige of the Soviet Union.

None of Russia's rockets developed to date could put a satellite into space, but the new ICBM would have the necessary power to put objects weighing several tonnes into Earth orbit. Yet, while the decision to build a super-rocket for sending atomic warheads almost halfway

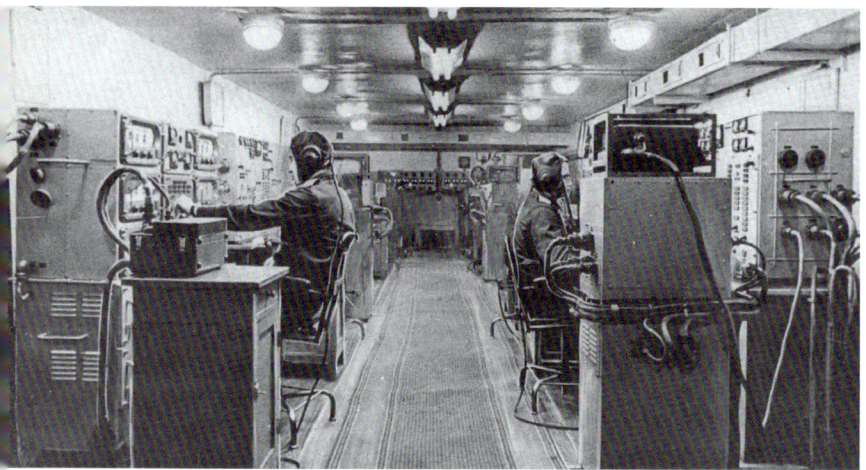

ABOVE Between 1945 and 1960 the Soviet Union recovered from a global conflict and forged ahead in the technological race to put satellites in space and send probes to the Moon. Much of this work was done with the simplest of radio, radar and very early computer engineering that belied their extraordinary success. *(TASS)*

BELOW Shot to an altitude of 132 miles (212km), a container launched by an R-2A returns to Earth by parachute carrying two dogs, Ryzhaya and Damka. *(TASS)*

across the globe was a continuation of the plan started by Stalin, the desire on the part of some scientists to pursue space research as a future goal was kept in the background – for now. In Soviet Russia it was all too easy to be accused, even falsely, of working against the interests of the Party and the State, as Sergei Korolev had discovered to his cost.

Already by 1953, impending developments in air defence, anti-aircraft missiles and electronic warfare were challenging the concept of the survivable long-range bomber. It became increasingly apparent to the Soviet leadership that their military interests would be better served by ballistic rockets. An ICBM could fly 200 miles (322km) into space, high above unfriendly countries encircling the USSR, and descend to its target at speeds in excess of 10,000mph (32,200kph).

The design of the world's first ICBM, designated the R-7, was finalised in late 1953 and authorised by order of the Council of Ministers on 20 May 1954. Seven weeks later, on 6 July, the project was raised in status to one of 'national importance'. The brainchild of Sergei Korolev, the R-7 was unlike any missile before it. To achieve the goal of throwing a three-tonne mass a distance of 5,280 miles (8,500km) would require much more powerful rocket motors ignited on the ground without the complexity of igniting an upper stage in flight.

There were other challenges too. Serious problems in developing rocket motors with high thrust levels forced engine designers to adopt a unique method of getting high thrust from a single engine. They did this by clustering four combustion chambers in a square, fed with propellants through a single turbopump and a single set of lines for fuel and oxidiser. Each quad cluster would be attached to the base of four separate boosters, surrounding a central core with its own unique quad cluster.

With all rocket chambers firing at lift-off the power required could be spread across 20 combustion chambers, each of which had an achievable thrust level, avoiding the need for a single large motor of exceptionally high thrust or an upper stage that would be required to ignite at altitude. The four outer boosters would be jettisoned on the way up, leaving the central core stage to continue firing, delivering a nuclear warhead to its target – or a satellite into orbit.

With the R-7 ICBM, Korolev had the perfect tool for realising his dream of space flight. It was a dream also held by the German rocket team led by Wernher von Braun, now working in the US on missiles that would usher in the space age along with Korolev's rockets. In fact, with its liquid propellant oxygen/hydrogen propellants that would take several hours to prepare for flight, the R-7 was much better suited to the role of space booster than as a missile, which would have to be ready to fire at very short notice.

Dreams of big things

Since 1956 the Russians had been preparing what they believed would be their first satellite. Not PS-1 or PS-2, but the much grander Object D. However, technical delays built up and erased any hope of its launch before 1958 at the earliest, so, to ensure they beat the Americans, the two smaller satellites were launched first. The third satellite, now designated Sputnik 3 but still known internally as Object D, was finally launched on 15 May 1958. With a weight of 3,930lb (1,327kg), it carried a suite of 12 science instruments in a conical structure with a height of 11.7ft (3.57m) and a maximum width of 5.68ft (1.73m).

By the time it was launched, Object D was following in the wake of three successful US satellites – one in January and two in March 1958 – and three failures. But the Russians too were having bad luck. Object D#1 was destroyed on 27 April when its R-7 rocket blew up, but its backup, D#2, was launched successfully less than three weeks later.

At more than 1.3 tonnes, Object D (Sputnik 3) had shocked the West. With this capability, delivery of a 3-tonne warhead to a target in the United States was possible, a weight which was greater than that required for an atomic bomb. Unaware of the depth and breadth of Soviet rocket engineering, to the West the Russians suddenly had the capability of hitting the US without warning. But it was the product of many

BELOW Soviet premier Nikita Khrushchev inherited the rocket programmes begun by Stalin and presided over an expansion of this technology into the R-7 and subsequent derivations for Russia's expanding space programme of the late 1950s. *(Novosti)*

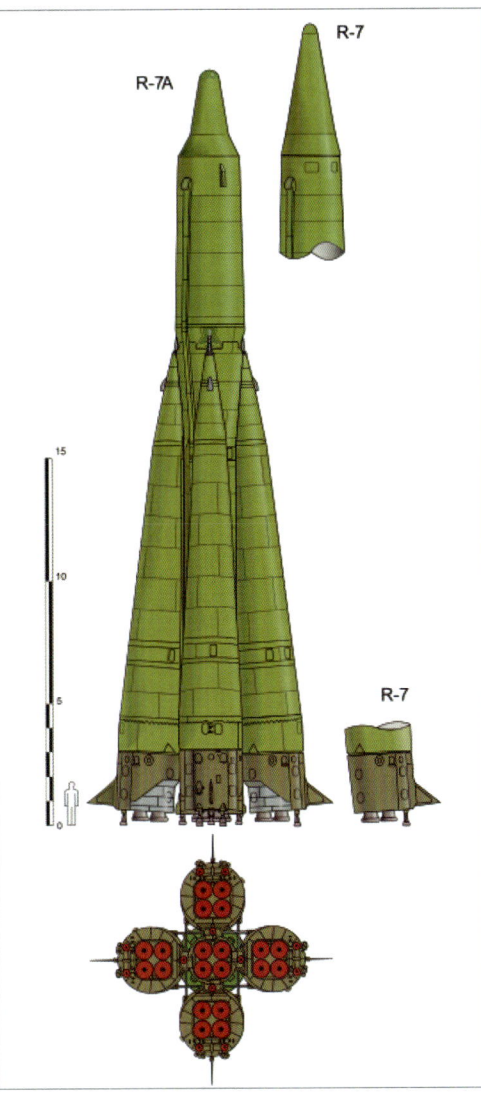

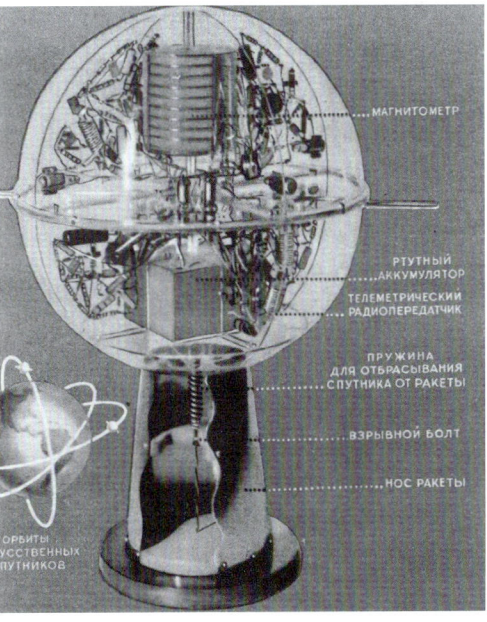

LEFT The R-7 intercontinental ballistic missile was developed by Korolev's team into a useful satellite launcher. A core stage was assisted by four boosters, all five elements having the same design of rocket motor, each with four combustion chambers. *(David Baker)*

ABOVE This special postage stamp was released to help celebrate the launch of Sputnik 1 and promote its propaganda value. *(David Baker)*

LEFT Finding the Russian people to have an insatiable appetite for its detail, the Russian magazine *Young Technology* displays the range of equipment inside Sputnik 1. A magnetometer is displayed at the top with an accumulator, a radio transmitter, and explosive bolts for separating the satellite from the rocket. *(David Baker)*

ABOVE Professor Blagonravov (left), A.M. Kasatkin and Sergei M. Poloskov of the Soviet Academy of Sciences delight in Russia's achievement. *(TASS)*

ABOVE Astrophysics students at the Institute of Geodesy in Novosibirsk watch the flight of Sputnik 1 through special reflecting telescopes on 7 October, signalling their observations down the line to the next group along its flight path. *(Associated Press)*

LEFT Andrew Ledwith of the Smithsonian Astrophysical Observatory in Cambridge, Massachusetts, explains the chart recordings of Sputnik 1's bleeping signal as it passed over Boston at 8:00am local time on 7 October 1957. *(David Baker)*

BELOW Press telephone traffic supervisors Les Davidson (left) and Dave Yelland tend receivers at the Associated Press newsroom in London on 5 October 1957 as they pick up signals from Sputnik. *(David Baker)*

RIGHT Britain's Astronomer Royal Richard Woolley poses by a globe of the Earth at a special gathering at the Royal Society in Burlington House, Piccadilly, London, convened on 10 October 1957 to answer questions from the press about Sputnik 1. *(David Baker)*

LEFT Dr Homer E. Newell (left) of the US Naval Research Laboratory and a member of the US satellite effort for the International Geophysical Year discuss Russia's satellite with John Townsend (centre), a member of the IGY rocket panel, and Dr Richard W. Porter of the technical panel on America's Vanguard satellite programme. *(David Baker)*

ABOVE Professor Harrie S. Massey (left), head of a British delegation to a UN conference on the International Geophysical Year, and William T. Blackband, also a British delegate, share the jubilation over Sputnik 1 with Professor Sergei Poloskov at the Russian Embassy in Washington DC. *(David Baker)*

RIGHT Wireless services feeding the press were quick to pick up signals from Sputnik 1 as it passed overhead. San Francisco newsman Bill Faulkerson takes a tape feed while an assistant measures the time lag on the fading signal. *(David Baker)*

BELOW An oscillograph at the Netherlands Postal Service receiving station at Nederhorst den Berg shows a strong signal coming in live from Sputnik 1 on 8 October 1957. *(David Baker)*

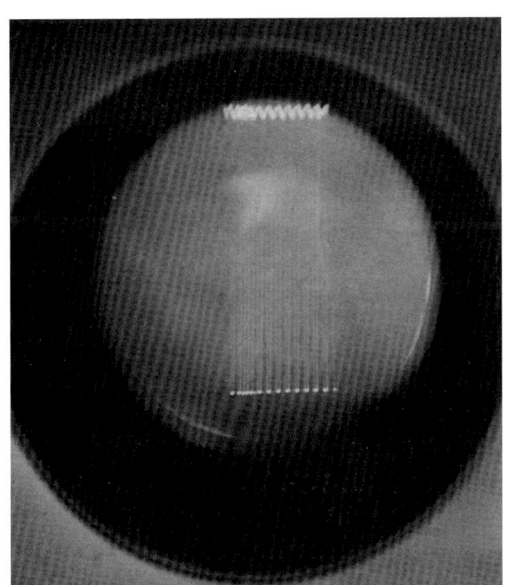

11
A WORLD LEAD

RIGHT A Soviet R-7 ballistic missile is adapted to launch Sputnik 2, which carried the dog Laika into orbit on 3 November 1957, and which remained attached to the core stage of the rocket. It burned up on re-entering the atmosphere on 14 April 1958, long after Laika had died. *(Boris Chertok)*

years of secret development work, and Sputnik 3 represented the heaviest payload the basic R-7 could put into orbit.

Paradoxically, and because the Russian scientists had never realised the euphoria that would surround their first satellites, they were unprepared for the demands from the Kremlin for ever more dramatic and spectacular flights. While Western intelligence officials scrambled to gather scraps of information about the R-7 – not even its shape or size was known – the Russians began to think about bigger things. Putting a man in orbit and sending probes to the Moon and the nearer planets came top of their list.

Since 1955 Korolev's design bureau, OKB-1, had been working on several optional designs for boosting a pilot into space on a brief suborbital flight using an early military rocket, the R-2A. Little more than a stretched development of the V-2, it was not powerful enough to put a man in orbit. Nevertheless, the R-2A was ready by 16 May 1957, on which date it sent two dogs, Ryzhaya and Damka, on a brief flight to a height of 132 miles (212km), experiencing six minutes of weightlessness before they were safely recovered.

This flight, following just a day after an unsuccessful attempt to launch the first R-7 ICBM, was the first in a series that was expected to result in the flight of a human occupant. Two months earlier, on 8 March, Korolev had gathered together a group of 30

LEFT A Russian postage stamp issued for Sputnik 2, the first flight of a live animal into orbit. *(David Baker)*

BELOW Young seven-year-old John Boardman is lifted up by one of his friends to see the replica of Sputnik 2 which was on show at the 'A Peep into the Future' display in the 'Earth as a Planet' exhibition in Gillingham, Kent, during November 1958. *(David Baker)*

BELOW A replica of Sputnik 2 is put on display in the Russian pavilion of the Brussels World Fair in April 1958. *(David Baker)*

RIGHT Object D, known as Sputnik 3, was to have been the first satellite launched, but delays brought forward the flight of the much smaller and simpler Sputnik 1. At more than 1.3 tonnes, Sputnik 3 realised the full lifting potential of the basic R-7. *(David Baker)*

young engineers and tasked them with planning for a progressive space programme. It was proposed fully eight months before the launch of Sputnik 1, such was his confidence in its success and his ability to get the approval of the Politburo for a concerted programme leading to manned orbital flight.

Working under Korolev, planning actually began in November 1956 when Mikhail Tikhonravov proposed a phased programme which included a manned capsule and unmanned flights to the Moon. These would emerge, respectively, as the Vostok spacecraft and the Lunik probes. Yet, while the Americans forged ahead with setting up the civilian space agency NASA, rapidly setting out an ambitious plan across all aspects of space research and exploration, the Russians had no politically approved central plan and little structure.

OKB-1 was funded by the military, and Tikhonravov's initial proposal was for a capsule that could be used for carrying a spy camera (OD-1) and a different version of the same design (OD-2) capable of carrying a dog and other biological experiments. By July 1957, with the launch of their first satellites still several months away, Korolev and Tikhonravov received tacit approval to develop the OD-1 spy satellite, with the Vavilov State Optical Institute developing the cameras.

There was less enthusiasm for developing the biological OD-2 satellite, for which the military saw little use. At this time the explosion of acclaim that surrounded the flight of Sputnik 1 had yet to erupt and the propaganda value was not a consideration. But each design was similar: a ball-shaped re-entry capsule with an equipment section comprising a cone-shaped structure attached to the front. The OD-1 re-entry capsule would contain a camera and film cassettes for recovery and processing of the film; the OD-2 capsule could carry a human occupant.

The two designers were devious in their

BELOW A Russian diagram of the scientific satellite Sputnik 3 launched on 15 May 1958. With a length of 11.7ft and a width of 5.68ft, it carried extensive scientific instruments for measuring the radiation environment of the Earth. It followed a failed launch on 27 April 1958, when the rocket blew up. *(David Baker)*

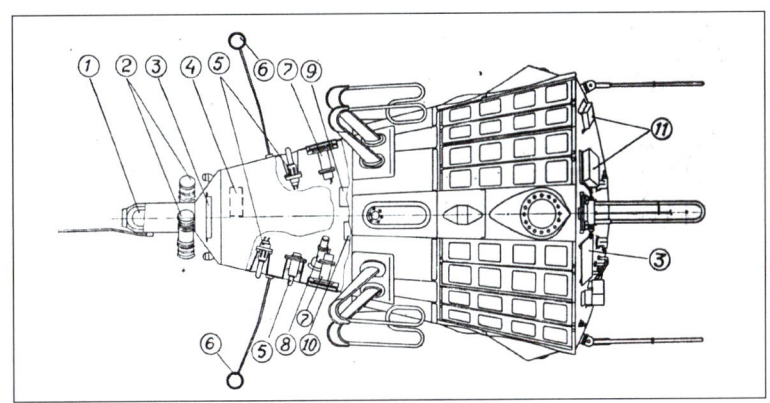

BELOW The replica of Sputnik 3 was used extensively for publicity purposes as the Soviet government showed with pride the achievements of Russian space scientists. *(David Baker)*

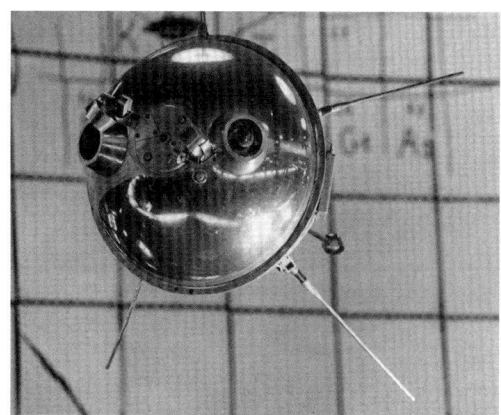

RIGHT Launched on 2 January 1959, Luna 1 was Russia's first space probe to successfully fly past the Moon and enter heliocentric orbit. *(David Baker)*

ABOVE Lida Volovnik shows off a replica of Luna 2 that was on display at the Soviet Trade Fair which ran at Earls Court, London, between 7–29 July 1961. Luna 2 was launched on 12 September 1959 and was the first man-made object to strike the Moon, reaching the surface on 14 September at 9:02pm Universal Time. *(David Baker)*

LEFT A model of the Luna 3 spacecraft launched on 4 October 1959 displayed at the Soviet Academy of Sciences Pavilion at the USSR Economic Achievements Exhibition in Moscow during May 1960. *(L. Velikzhanin)*

use of a common design, converging different operational applications into a common vehicle. By pressurising the re-entry capsule with an oxygen/nitrogen sea-level atmosphere, the military could avoid the expense of developing special cameras to work in a vacuum. And by providing a breathable atmosphere, the same basic design could carry living things into orbit.

But humans in space was the fervent dream of OKB-1, and Korolev brought in Konstantin Feoktistov to head-up a special engineering section to push forward the design of a man-carrying capsule. A range of other departments and different institutions were drawn in to the effort and the original OD-2 design was retained for a broadly expanded concept that had matured by August 1958 into a practical design.

Within six months OD-2 had changed dramatically. The instrument section providing electrical power, communications and attitude control, plus oxygen and nitrogen bottles for pressurisation, would be in a double-cone-shaped structure below the spherical re-entry module. Also, the OD-1 spy satellite design was abandoned for a variation of the OD-2 design, thus simplifying further the adoption of a single spacecraft for two very different roles. Korolev was fighting against hesitancy on the part of the military, fearful that funding for the spy satellite would be compromised by ideas of human space flight.

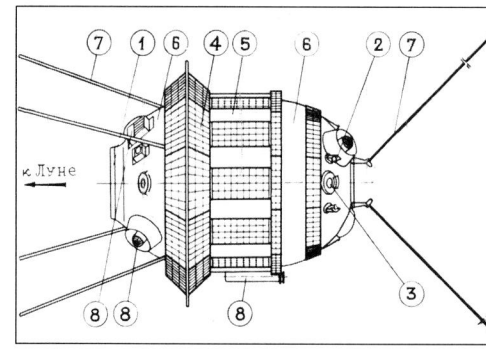

RIGHT Luna 3 weighed 614lb (278.5kg) and was launched by an 8K72 rocket, with an upper stage from which would be developed the rocket that would put the first man in orbit. *(David Baker)*

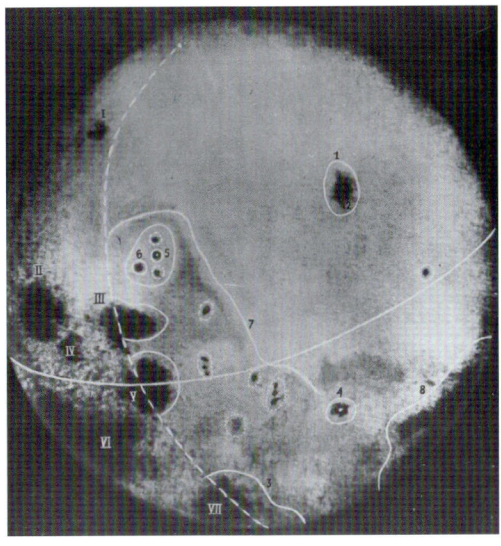

ABOVE Luna 3 took this photograph of the far side of the Moon after it had sped past: because it continually faces away from Earth, this had never been seen before. Luna 3 was the first spacecraft to be attitude-stabilised and took 29 pictures on 7 October 1959, of which at least 12 were successfully transmitted back to Earth. *(David Baker)*

BELOW While Russia was basking in the glorious achievements of its successful space programme, the Americans were adapting small military theatre missiles for satellite flights. This Redstone was the progeny of the US Army team with Wernher von Braun as technical director, and would be hurriedly adapted for placing America's first satellite in orbit on 31 January 1958. *(US Army)*

Meanwhile, Russia seized on the furore caused by the first three Sputniks and began a series of flights to the vicinity of the Moon. Only the successful ones were publicly announced and numbered in sequence. On 2 January 1959 Luna 1 was sent beyond the Moon, the first object to reach escape velocity. Before the end of 1959, Luna 2 became the first probe to impact the Moon, followed by Luna 3, which shot past our nearest celestial neighbour and transmitted to Earth a fuzzy picture of the Moon's far side.

This was the first time anyone had seen what the other side of the Moon looked like, further enhancing the mystique surrounding the top-secret Soviet space scientists and engineers.

LEFT The one enduring consequence of the Soviet space achievements was the urge to establish a central organisation for conducting America's non-military space programmes, that being the National Aeronautics and Space Administration (NASA), which opened for business on 1 October 1958, T. Keith Glennan its first Administrator. *(NASA)*

LEFT President Eisenhower insisted that the US contribution to the International Geophysical Year would be Vanguard, ostensibly a civilian programme using a rocket developed from several existing motors. It languished within a half-hearted effort that did not see a successful satellite launch under this programme until 17 March 1958. *(NASA)*

Chapter Two

Vostok and Voskhod

Probably as a result of America's one-man Mercury programme publicly announced in October 1958, but at the urging of Korolev and with the tacit support of Khrushchev, on 5 January 1959 the Russian government gave its formal approval for the start of its own manned space programme.

OPPOSITE The basic design of the Vostok spacecraft consisted of a spherical re-entry module attached to an equipment section and the second stage of the 8K72K rocket, which put it into orbit. Note the restraining straps on the re-entry modules and the radio communication antennae. *(David Baker)*

ABOVE The design and sizing of the Vostok spacecraft was based on the lifting capabilities of the 8K72K rocket, itself developed from the R-7 ballistic missile but with a Block E second stage and an RD-0109 developed from the stage utilised for lunar flights. *(David Baker)*

When the OD-1 module design merged with OD-2, Korolev changed the designation to Object K (*Korabl*, the Russian word for 'ship'), of which he proposed four versions: 1K as a prototype for both spy satellite and manned spacecraft; 2K and 4K for operational spy satellites; and 3K for the manned variant. Korolev signed off on the design of these on 17 March 1959, and on 22 May the Russian government formally approved what would become the Vostok spacecraft.

Object K (Vostok) was now to be a spacecraft of two parts: the spherical descent module and the double-cone-shaped equipment module. The descent module had a diameter of 7.5ft (2.3m) and a total all-up weight of 5,425lb (2,460kg) and was capable of carrying a single occupant on an ejection seat inclined slightly to the horizontal. This provided a means of escape during the time the spacecraft was on the launch pad or during the first few seconds of ascent. Korolev had tried to integrate a launch

RIGHT His bust displayed at the cosmonaut training centre outside Moscow, Sergei Korolev fed the increasingly insatiable appetite of Nikita Khrushchev for spectacular space flights and his feverish efforts to put the first man in orbit. *(David Baker)*

BELOW Support for development of the Vostok spacecraft was only obtained from the military by Korolev's proposition for the Zenit spy satellite, a copy of the manned space vehicle but with cameras and viewing ports instead of a pilot and window. *(David Baker)*

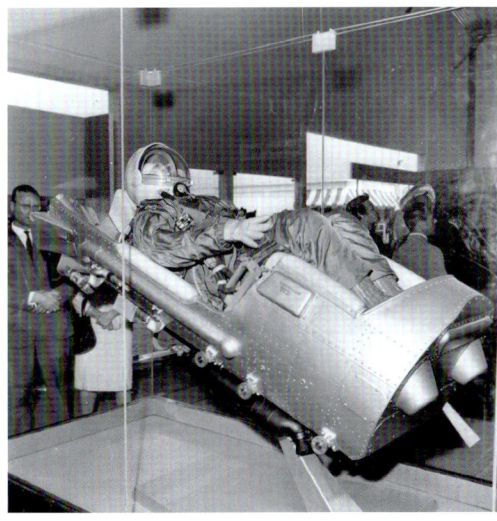

FAR LEFT The ejection seat was fitted to allow the cosmonaut to leave the re-entry module on the way down through the atmosphere, thus avoiding a bone-jarring touchdown. The cosmonaut could also use this as means of escape during ascent. *(David Baker)*

LEFT The underside of the ejection seat shows the piston cartridge, the entire seat design being lifted from a similar seat used on MiG fighters of the period. *(David Baker)*

escape system incorporating a tower design that would fire to pull the spacecraft free of the rocket in the event of an abort, but it proved too heavy for the launch vehicle.

The thick heat shield comprised a thermal protective layer covering the complete exterior that weighed 1,850lb (837kg). The seat itself weighed 740lb (336kg) and was designed to eject, along with its occupant, through a circular hatch in the side of the module at an altitude of 23,000ft (7,000m), with parachute deployment at 8,200ft (2,500m) lowering the module to the ground. The cosmonaut could elect to eject during descent and return to Earth by parachute or ride in the capsule down to the ground, risking a very heavy, bone-shaking landing!

The descent module itself was pressurised at 14.7lb/in^2 (761mm Hg) with a mixture of 79% nitrogen and 21% oxygen, closely resembling the atmosphere at the surface of the Earth. The environment inside the module was reconstituted with a chemical bed of alkali metal superoxides that reacts with the exhaled carbon dioxide to produce oxygen. A separate heat exchanger would maintain an acceptable temperature inside the descent module, with a cooling and desiccation unit controlling humidity. Much of the technology for the environment

ABOVE The spherical design of the re-entry module was balanced so that it would right itself on entering the atmosphere with the cosmonaut lying on his back during deceleration, thus being in a better position to withstand the g forces. *(David Baker)*

RIGHT The re-entry module was attached to the instrument section with four straps, and power and conduits for communication and emergency air supply were carried by a spoon-shaped umbilical linking the two modules. *(David Baker)*

LEFT Status indicator lights and analogue dials provide basic information for the cosmonaut, who would report conditions aboard the spacecraft. Engineers on the ground relied on voice as well as telemetry to provide data on the status of the various systems in the brief periods of each orbit when it was within range of Soviet tracking facilities. Vostok had four switches and 35 indicators; NASA's Mercury capsule had 56 switches and 76 indicators. *(David Baker)*

in Vostok had been developed through experiments with animals on the smaller ballistic rockets in the early and mid-1950s.

The separate equipment module consisted of two aluminium half-cones attached together at their widest point with a maximum diameter of 7.9ft (2.43m) and a length of 7.4ft (2.25m). When loaded it had a maximum weight of 5,005lb (2,270kg), packed with essential systems necessary for supporting the descent module throughout its mission in a pressurised environment similar to that in the descent module. The equipment module had two separate systems each with eight cold-gas nitrogen attitude control thrusters and 44lb (20kg) of propellant. Each thruster delivered 3.3lb (15N) of energy. Batteries provided about 24kW/hr of electrical power, with an average power drawdown of 200W.

The main retro-engine was known as the TDU-1 and occupied most of the equipment module interior, with 606lb (275kg) of propellants consisting of a mixture of red-fuming nitric acid (RFNA) and amine. RFNA is the oxidiser and consists of 84% nitric acid, while amine is the fuel which when injected into the combustion chamber in the presence of RFNA ignites spontaneously. Highly toxic and corrosive, these propellants were stable, storable and would ignite without the need for a complex ignition system. TDU-1 had a thrust of 3,560lb (15.83kN) for 42 seconds, reducing the speed of the spacecraft by 350mph (560kph) and starting it on a path down towards the atmosphere.

LEFT Cosmonaut Vladimir Komarov's workbook with personal notes on the manual controls in Vostok. *(David Baker)*

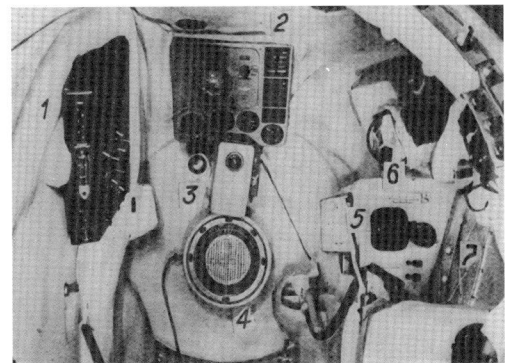

ABOVE The basic layout of the cabin was spartan by modern standards but the design emphasis was one of automated control, there being little that the cosmonaut could do to 'fly' the spacecraft. *(David Baker)*

LEFT The control panel was situated on the left side of the cosmonaut, with switch positions for directing electrical power to various systems. *(David Baker)*

The descent and equipment modules were attached with straps and would remain a single integrated unit until shortly before re-entry, when the TDU-1 would fire for the de-orbit burn. Only then would it separate, burning up behind the descent module and its protective heat shield. Several attitude and orientation systems were provided, since correct orientation prior to the retro-burn was essential for getting the descent module on its correct flight path. The primary system was a Sun sensor, with an infrared sensor for locating the Earth as a backup attached to a complex system of gyroscopes.

The cosmonaut had a backup mode that allowed him or her to manually fire the retro-rocket in the event of a failure to the primary or secondary systems. For this situation, the cosmonaut would use a Vzor optical device, a circular window on the opposite side of the spacecraft with alignment marks for controlling yaw and a mirror device providing visual cues for pitch and roll. In this way the pilot would place the spacecraft roughly in the correct attitude for manually firing the TDU-1 retro-rocket. But this was the only concession to the cosmonaut, otherwise riding purely as a passenger.

The Vostok spacecraft was designed as a system totally controlled by electronic signals sent from the ground. The occupant was not expected to play any part in deciding, or controlling, his flight other than manually operating a very few emergency procedures

BELOW A Lockheed U-2 spy plane took this view of the launch facility where Sputnik 1 was launched and from where Vostok cosmonauts would fly. Looking directly down from high altitude, the flame trench fans out from beyond the cantilevered launch stand. *(CIA)*

RIGHT The Vostok spacecraft positioned inside the shroud which protects it from atmospheric pressure during ascent, a circular cut-out provided for use of the ejection seat should that be necessary before the shroud itself was ejected. *(David Baker)*

critical to saving his life. The ability to manually override automatic or ground-commanded systems was reserved for only dire situations, and even the codes to operate those controls were sealed until opened on orders from the control centre.

There were sound reasons for this in that the physiological and psychological reaction of the cosmonaut had yet to be understood – nobody had yet been into space. But there were other reasons too. Centralised control was an integral element in Soviet thinking at the time, and the design engineers were determined to control the missions themselves from start to finish. They had designed and built the spacecraft and they were unable to accept that a pilot had better knowledge or experience than them in ensuring the satisfactory outcome of the flight.

This was a fundamental difference between the US and Russian approaches to space flight which would dominate their thinking for several decades, and play a major role in how the Soyuz spacecraft was designed. It was an issue that had already been worked through in the US, where rocket engineers such as Wernher von Braun believed that pilots could not operate satisfactorily in space and that only the rocket engineers should have control of manned space vehicles, whereas the pilots had a different view and had fought hard to retain as much manual capability as prudently possible.

Vostok was completely different to NASA's Mercury spacecraft. For one thing, the tiny Mercury had a diameter of only 6.2ft (1.8m), a length of 6.8ft (2m), a weight on orbit of

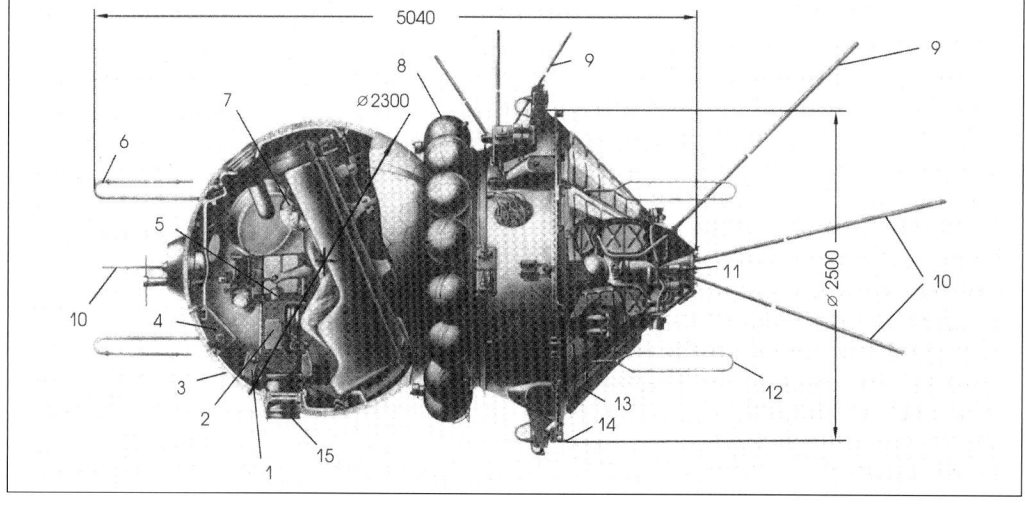

RIGHT In space, the second stage is jettisoned and the spacecraft operates on the systems contained in the equipment section behind the re-entry module. *(David Baker)*

3,000lb (1,400kg) and an internal volume of 100ft^3 (2.8m^3). The Mercury capsule was engineered with multiple redundant attitude control systems designed to be operated by the astronaut. Moreover, the pure oxygen pressurised environment was held at about 5lb/in^2 (258mm Hg), approximately one-third sea-level atmospheric pressure. A drawback with the Vostok system was the danger of the cosmonaut getting the 'bends' if the internal pressure failed, the pure oxygen of the pressure suit becoming their sole means of survival.

Together, the Vostok descent module and the equipment module weighed more than 10,400lb (4,700kg) – more than three times NASA's Mercury spacecraft, and three times the weight that the R-7 rocket could lift into orbit. To do that, a new upper stage was needed. The factory designation of the R-7 used for launching the first satellites into space was 8K71 and a series of modifications were made subsequent to that for more ambitious space missions. The first, carrying an upper stage developed to send the first Luna probes in the direction of the Moon, was known as the 8K72. It was a developed version of this, known as the 8K72K (the 'K' again indicating *Korabl*), which was developed for manned Vostok flights.

The upper stage was known as the Block E, with an engine designated RD-0109 which had a thrust of 12,260lb (54.5kN), a little more powerful than the upper stage and its RD-0105 engine used for the Moon shots. Together, the Block E stage and Vostok had a length of 28.5ft (8.7m) and the stage had a diameter of 8.5ft (2.59m). The first launch of this combination took place on 22 December 1960, but the upper stage cut off too early and the unmanned Vostok fell back into the atmosphere. This flight had been preceded between 15 May and 1 December 1960 by four attempted flights of similarly unmanned and lighter versions of Vostok using the 8K72 vehicle with its slightly less powerful upper stage.

Of those four flights the first failed when the spacecraft pointed in the wrong direction during retrofire and was propelled into an orbit from which it would not re-enter the Earth's atmosphere for five years! Another, carrying two dogs named Chaika and Lisichka failed at launch; one carrying dogs Pchelka and Mushka was

LEFT With little space, some cosmonauts unstrapped themselves and floated free, experiencing extended weightlessness for the first time. *(David Baker)*

destroyed during re-entry; and another carrying the dogs Belka and Strelka was a total success.

Man in space

No sooner had the Soviet government given its approval to the development of a manned spacecraft than preparations began for selection of a suitable list of cosmonauts. By the end of 1959 the Air Force had short-listed 20 candidates, which unlike their US counterparts were not required to have extensive flying experience. They were considered passengers and little more. Instead, fitness, mental discipline, intelligence and the ability to handle physical stress were paramount.

On 30 May 1960 the six short-listed finalists were selected: Gagarin, Katashov, Nikolayev, Popovich, Titov and Varlamov. At the end of June the Air Force established the cosmonaut training centre at the Zelenyy suburb about 20 miles (30km) north-west of Moscow, later known famously as Zvezdniy Gorodok. Katashov and Varlimov were eventually eliminated for physical reasons and replaced with Bykovsky and Nelyubov.

The first Vostok spacecraft was designated 1KP, the 'P' denoting it was a test of basic systems and, because it did not carry a heat shield, would not be recovered. The official designation of Korabl-Sputnik 1 announced to the outside world was intended to allay suspicions in the West that this was a completely unique vehicle, designed to carry a

ABOVE Andrian Nikolayev in training, which involved familiarisation with all aspects of the mission including survival in remote areas where there was a strong possibility the re-entry module might land. *(David Baker)*

BELOW Nikolayev, briefly simulating weightlessness for less than a minute during a parabolic aircraft flight. *(TASS)*

cosmonaut. It was launched on 15 May 1960 by an 8K72 rocket utilising the RD-0105 stage previously used for the Moon shots. Tests were successfully carried out for nearly four days and the retro-motor was fired, but problems with the orientation system had the spacecraft pointing at an incorrect angle and it ended up in a higher orbit. It would not re-enter the Earth's atmosphere until 15 October 1965.

The second flight took place on 28 July 1960 carrying the dogs Chaika and Lisichka, but a malfunction in the rocket caused an explosion just 28 seconds into flight, although the flight was never acknowledged by the Soviets at the time. Although the re-entry module separated from the stack the two dogs were killed outright by the force of the explosion.

The third flight of a Vostok spacecraft (Korabl-Sputnik 2), launched on 19 August, carried two more dogs, named Belka and Strelka, which were successfully placed in orbit. The spacecraft carried 12 mice and a wide range of biological specimens for testing cellular reactions to space, and two TV cameras transmitted views of the dogs in space. News of this success, heralded by Russia as the second Korabl Sputnik carrying the first living animals to be returned from orbit, was brought to the Halls of Columns in Moscow, where Soviet judges were considering the sentence for U-2 spy-plane pilot Francis Gary Powers!

To the outside world, there was no official link made between these flights carrying dogs and biological specimens and any attempt to get a man into space, although the obvious link was not lost on NASA's Mercury team. However, within the Soviet Union the increasing momentum attracted greater attention from the upper levels of the Ministry of Defence and of the government too, and the successful flight of Belka and Strelka catalysed that scrutiny. In a striking memorandum issued by Korolev on 7 September he enumerated the many separate developments taking place within the Vostok programme but gave a warning that had a great influence on how the space programme was seen by external bodies.

As we shall see later, the development of a successor to Vostok (Soyuz) was already on the drawing boards and Korolev knew that a transformation in the way the industry was run would have to take place if success were to become a byword for space operations. The fundamentals would plague the space programme down through the decades, but they were characterised by Korolev in ways which would influence the future of Russia's human space-flight programmes.

Writing essentially for the eyes of military men hardened by acceptable losses of lives on the battlefield, Korolev said that the old way of building military equipment, be it tanks or rockets, would not do for the space programme. It was not acceptable, he said, to be satisfied with a level of reliability that tolerated the occasional failure. That might work as the price of a compromise between the exacting engineering requirements of sophisticated war machines, and the need to produce equipment handled by recruits and non-specialists on the battlefield where the occasional missile misfiring was acceptable; but it would not work for spacecraft.

With space equipment, said Korolev, the total numbers flown were so few, with so much riding on each flight, that failure rates commonly accepted for battlefield equipment would bring disastrous consequences for the lives of cosmonauts if accepted for space. A new attitude was needed, one which had no place for accidents or failure. Korolev reminded his audience that the Vostok spacecraft and its launcher had six prime propulsion elements, with 33 structurally loaded systems involving high pressures and thermal tolerances. The spacecraft itself had 241 vacuum tubes, 850 connectors each with up to 850 contacts, more than 6,000 transistors, 56 electric motors, 800 relays and switches and more than 9 miles (15km) of cabling.

Korolev was not the only space engineer to see that a revolution in manufacturing, quality control and testing was essential to the success of the human flight programme. He secured 11 top-level signatures from engineer-managers in 123 departments covering 36 plants in what he himself saw as a reminder that the entire programme now had to move up a notch to deliver as promised an outstanding propaganda tool for the Soviet Union. But

ABOVE Dogs were used to pave the way for men, several canine flights testing the systems and the physical reaction of the animals to orbital space flight. *(Novosti)*

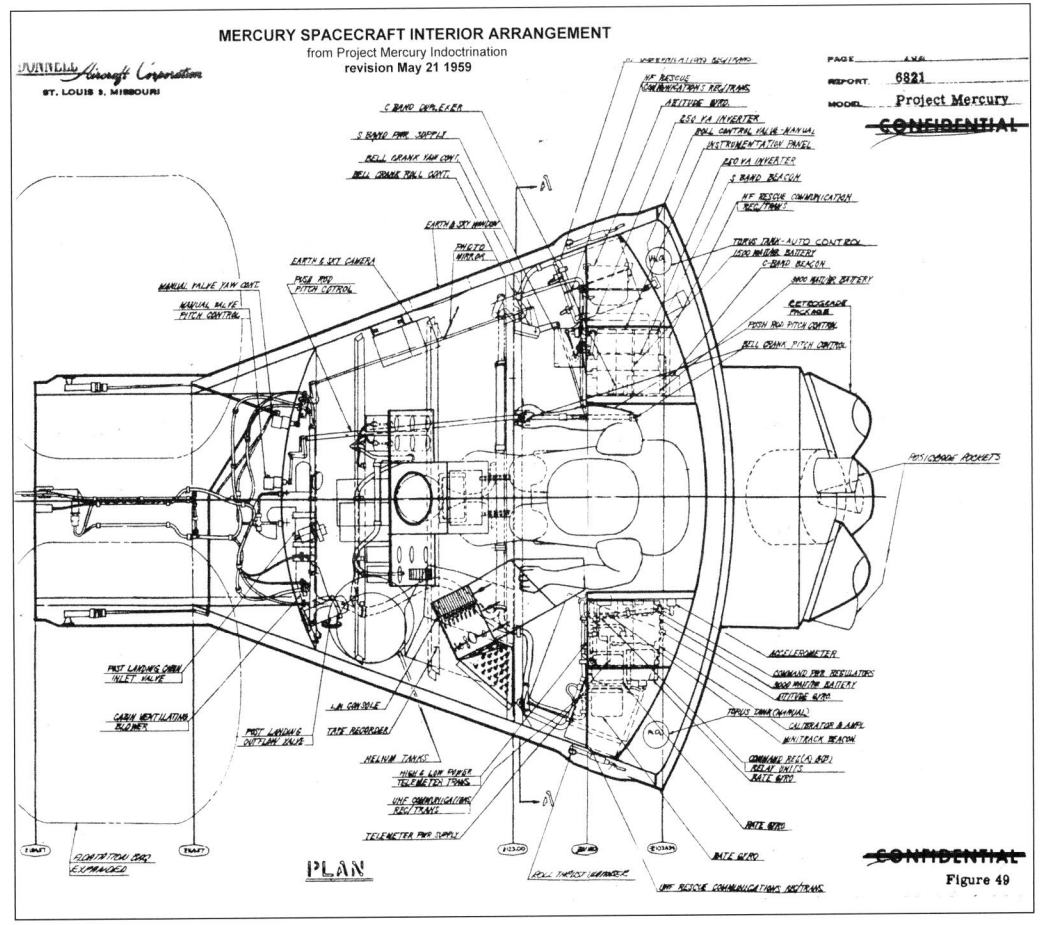

LEFT America's Mercury spacecraft was smaller than Vostok, and while, like Vostok, it could not change its orbit it did provide greater manual authority over attitude and the control of spacecraft systems. *(NASA)*

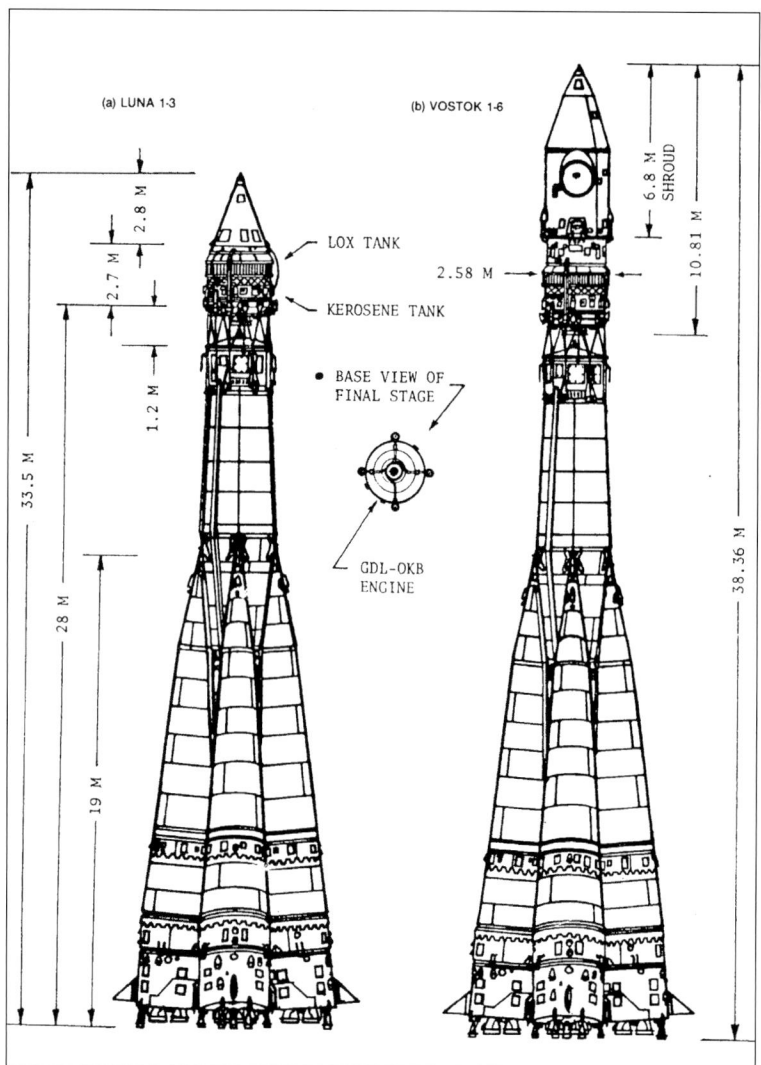

ABOVE Derivatives of the R-7 were to provide the means for Russia to launch space probes to the Moon and send cosmonauts into orbit. At left is the 8K72 with the upper stage used for sending probes to the Moon, compared with the 8K72K for Vostok flights, the upper stage here being extended in capability. *(Charles P. Vick)*

for Korolev the politics were an irritating but necessary side issue, essential nevertheless to getting approval and top-ranking support for his uncompromising and demanding approach. As a memo to the initiated it reminded them what most of them knew already. To the external audience it was a wake-up call.

The September memorandum was crucial to the success of Vostok but also for setting the pattern of management and control necessary for later successes, including Soyuz, as plagued as they would be with their own failures. It also triggered a formal declaration from the Vostok team to the Central Committee of the Communist Party outlining the remaining flights it deemed necessary to achieve the first manned flight, claiming that this could be accomplished within two months. Prior to that, two further flights with the precursor Vostok 1K spacecraft would be needed, followed by two automated flights with the definitive Soyuz 3KA. With full approval granted, Korolev set about preparing the next four flights to fully qualify Vostok for its design role.

But on 24 October a disaster of epic proportions at the Baikonur launch facility brought a temporary halt to Vostok. An R-16 rocket designed and developed by Mikhail Yangel had erupted when the second stage suddenly ignited and burned fiercely, setting light to all the propellants in the first stage. The erupting fireball spread burning fuel and toxic fumes throughout the surrounding area, killing 126 people. Included in the dead was Marshal Mitrofan Nedelin, commander-in-chief of Soviet strategic rocket forces, identified only by his medals. Yangel had developed the R-16 as a two-stage missile using storable propellants, fuel and oxidiser that remain liquid at room temperature. Using liquid oxygen (LOX) as the oxidiser, Korolev's R-7 was not suitable as a quick-reaction weapon due to the amount of time it took to prepare and the limited amount of time the LOX could be held in the tanks.

The effect was to bring a two-week halt to preparations for Vostok flights, pushing back to February 1961 the first manned flight. Korolev and many of the industrial organisations working on his spacecraft had produced equipment for Yangel's R-16. It was quickly discovered that premature firing of the second stage had been caused by a stray electrical circuit that, in the absence of a safety breaker, had triggered the ignition command. If nothing else, it emphasised the reality of Korolev's memorandum.

Publicly announced as Korabl-Sputnik 3, and launched on 1 December 1960, the next flight carried dogs Pchelka and Mushka into orbit aboard the third Vostok 1K. They were to have been returned to Earth on the 17th revolution, but the TDU-1 engine misfired, not slowing the spacecraft sufficiently to bring it down as planned but instead taking the re-entry module another one and a half orbits around the Earth on its shallower trajectory. Unfortunately for the dogs, to prevent a capsule inadvertently falling to Earth outside the territories of the USSR it contained an explosive destruct package to prevent foreign powers getting a look at Russia's flagship spacecraft. The explosives were to be triggered if a timer set running at retrofire failed to sense the build-up of decelerating g-forces as planned, indicating a long descent path carrying the spacecraft far to the east. The explosives worked – but were removed for the remaining flights!

The next flight, the last of the four Vostok 1K types, should have carried dogs Kometa and Shutka into orbit on the first of the 8K72K rockets. Following launch on 22 December, the upgraded second stage shut down prematurely, resulting in the Vostok separating as planned. Thus propelled to a ballistic trajectory, it was carried to a height of 133 miles (214km) and sent 2,175 miles (3,500km) downrange from the launch site into the freezing Siberian wilderness. Urged on by a timer attached to another self-destruct system which would activate 60 hours after landing, rescue teams battled deep snow in temperatures of -40°F (-40°C) to reach the sphere, finding the timer had still not gone off.

Both circular hatches had blown but the dogs had not been ejected as planned, the seat slamming into the side of the porthole where it was lodged. Now more than 60 hours after it landed, the team had to disarm the explosives and extract the dogs – which were fit and well! While the dogs were flown back to Moscow, getting the capsule out was difficult. Hauling it across vast distances using makeshift transport, the team had to survive a night where temperatures fell to -76°F (-60°C), and it was not finally returned for two weeks.

With two failures in a row, albeit for different reasons, the February 1961 launch date for a human occupant was impossible. On 5 January the State Commission approved two more automated flights with the fully developed 3KA type, contingent upon which would rest the decision about a manned attempt. This time, each test flight would last one revolution of the Earth, precisely the duration planned for the first manned flight, and involve one dog carried in a special container adjacent to the ejection seat, which would be occupied by a mannequin.

Launched on 9 March 1961, the first carried the dog Chernushka in the second 3KA spacecraft on the new 8K72K launch vehicle (Sputnik 9, Korabl-Sputnik 4). Along with 80 mice, several guinea pigs, seeds, blood samples, human cancer cells and a host of other bacteriological samples, the dog survived its 1hr 46min flight as planned. Before the final test of an unmanned Vostok spacecraft, a tragic accident on 23 March took the life of Valentin Bondarenko, one of the 20 cosmonaut finalists selected by Korolev.

In a prescient warning of the dangers of an enhanced oxygen environment, he burned to death in a flash fire while locked inside a test chamber where the atmosphere contained 50% oxygen. Insulated against sound to simulate the psychological effects of isolation, he was ten days into a two-week simulation and stood little chance of survival. Less than six years later three NASA astronauts would die when their Apollo spacecraft ignited in a 100% oxygen environment.

On 25 March the final automated test mission took place when Zvezdochka followed her canine predecessors into orbit on a repeat of the previous flight. Designated Sputnik 10 (Korabl-Sputnik 5) it was a total success and released the next flight to a human occupant. But the snow and harsh conditions in the planned landing area prevented the recovery team from reaching the capsule and its zoo of animals. With freezing temperatures and deep snow, 24 hours had elapsed when they finally arrived at the site on horse-drawn sleighs, only to find that the mannequin, which had been ejected as planned, was the focus of concerned

BELOW Yuri Gagarin was selected from among six prospective candidates for flying in the first Vostok spacecraft to carry a human into space. *(Novosti)*

ABOVE The *Huntsville Times*, the local newspaper in the town where ex-V-2 designer Wernher von Braun was working for NASA as Director of the Marshall Space Flight Center building Saturn rockets, proclaims Russian prowess. *(NASA)*

attention from local peasants.

So intense were preparations for the first manned flight that Korolev decided to forego some pre-flight simulations. NASA's highly publicised Mercury programme would precede any attempt at human orbital flight with at least two ballistic shots propelling an astronaut straight up into space and down again in flights each lasting a mere 15 minutes. Plans to fly the first of these were nearing completion, and although they would pale in comparison to Korolev's massive Vostok and its flight around the Earth, to the general public it would represent a major coup. But the Americans were still many months away from their own astronauts flying in orbit. Nevertheless, the gap was narrowing and Korolev saw a successful Vostok flight as crucial to his entire plan for human space flight.

The successful flight of Yuri Gagarin aboard Vostok 1 on 12 April 1961 regained Soviet prestige. To the outside world it was a stunning demonstration of Soviet progress and of the technical sophistication of a people long thought by many to be backward and uneducated. To those privy to the mission, it was a success to be sure, but much remained to be done to refine the spacecraft further. Nevertheless, experiencing for the first time what it is like to return to Earth through the atmosphere at speeds greater than 17,000mph (27,350kph), Gagarin sampled a flight regime unknown to any other human at that date.

But all had not gone as planned. The equipment section remained hung up on loose cables and electrical lines after the straps holding it to the re-entry module were severed following retrofire. Only slowly did they burn through, allowing the spherical module to right itself for correct orientation. The hatch was jettisoned at an altitude of about 23,000ft (7,000m), followed two seconds later by the ejection seat and its occupant. The seat fell away at an altitude of 13,125ft (4,000m) leaving Gagarin to float down to Earth on his personal parachute. The parachute at the top of the descent module was released at 8,200ft (2,500m), lowering Vostok 1 to the ground nearby.

RIGHT The Russians produced booklets such as this English language edition to trumpet the great success of Yuri Gagarin. *(David Baker)*

In a single orbit of the Earth, a flight lasting 1hr 48min, Russia achieved a feat exceeding on several levels the first launch of an artificial Earth satellite just three years six months earlier. But Korolev, as well as the Soviet leadership, knew that in reality the country had a tenuous hold on leadership. Aroused beyond expectations, the Americans had risen to the challenge of Sputnik by investing in a wide range of satellite projects, and soon would launch their first spacecraft to a close fly-by of the planet Venus.

The Mercury programme was roughly equivalent to Vostok, albeit much smaller and with less capacity for remaining in space as long, but few in Russia expected the Americans to rest on their laurels. Challenged by Vostok 1, President Kennedy – in office for a mere 12 weeks – reacted emotionally to the coup, asking his Vice President, Lyndon Johnson, to find a way to beat the Russians. This set off a series of events within the government which persuaded Kennedy to make a Moon landing goal the challenge of the decade, pledging to get US astronauts on the lunar surface by the end of 1969.

Announced publicly on 25 May, less than three weeks after Alan B. Shepard flew the first ballistic Mercury flight, that declaration mobilised an effort that would eventually trigger reaction in the Soviet Union far beyond Korolev's expectations but well within his hopes. Korolev wanted to go to Mars; the Moon was considered a mere stepping stone. But in the meantime, Vostok could only do so much, and with the Americans pledging a fully funded effort with its Apollo programme, Korolev's thoughts turned to bigger and bolder missions as well as larger and more capable spacecraft.

Before then, more Vostok missions could be flown to fully exploit its capabilities and to get ahead of the Americans before they began flying orbital flights with their Mercury spacecraft. But to an astonishing degree

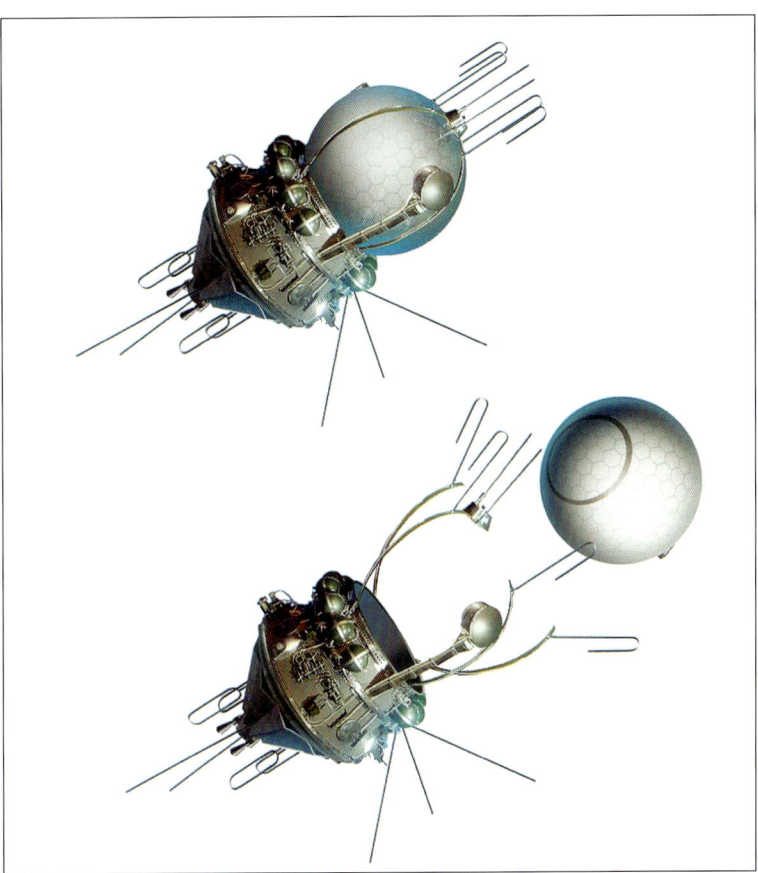

ABOVE The Vostok re-entry module was supposed to separate cleanly from the equipment module prior to re-entry but rarely did, the straps only fully burning through during descent. *(David Baker)*

FAR LEFT The Vostok 1 re-entry module is exhibited at the RKK Energia Museum in Moscow. *(David Baker)*

LEFT On landing, usually in an agricultural setting, the re-entry module could roll over or be dragged along by the parachute lines until released. *(Novosti)*

RIGHT A statue to Yuri Gagarin stands at Star City, the cosmonaut training facility outside Moscow. *(David Baker)*

BELOW Russia's first man in space, Yuri Gagarin (right) relaxes for a publicity photograph with Gerhman Titov, the second Russian to fly a Vostok and the first to spend more than a full day in space. *(Novosti)*

BELOW RIGHT Because flights were supposed to end with the cosmonaut ejecting before the re-entry module landed heavily, parachute training was a top priority, as practised here by Gerhman Titov. *(Novosti)*

there was no strategic blueprint for the programme, no road-map with clearly defined missions. The Vostok programme grew flight by flight, each mission designed according to the recommendations of Korolev or to accommodate some political objective. There was no pre-planned structure to how many missions would be flown, or when. At the time of Gagarin's flight 18 Vostok spacecraft were funded, half for human space flights and the other half for the spy satellite role.

Reaction to Gagarin's flight at home and abroad made it clear that further flights would follow and, with a new multi-person spacecraft on the drawing boards – later to mature into Soyuz – it was important to keep delivering new propaganda opportunities to the Soviet leadership. In a hotly debated discussion among senior Korolev managers, cosmonauts and others, the duration of the next flight was decided.

Earlier flights with the automated Vostok had demonstrated that the spacecraft could operate for a full day in orbit and this was what Korolev wanted, but the Air Force, which controlled the cosmonauts, was fearful of the ability of the human body to survive that long in space. But Korolev was a persuasive man and it was agreed that Vostok 2 should be targeted for a 17-orbit mission. Consequently, on 6 August 1961, Gerhman Titov was launched aboard Vostok 2 just over two weeks after Virgil 'Gus' Grissom flew the second ballistic Mercury flight.

Titov became the first man to sleep in space, returning to Earth, like Gagarin, by parachute after 25hr 11min. For much of the flight he had been extremely uncomfortable, complaining of nausea, disorientation, a stuffy head, and generally feeling unwell. He attempted to eat a three-course lunch from packaged tubes, and attempted simple tasks such as writing, but found the only moderate relief was to be had from sleep. At other times he took manual control of the spacecraft's orientation and was seen on TV live from the spacecraft, with 400 lines a picture much improved from the 100-line system used for Gagarin's flight.

During re-entry the equipment section failed to separate after retrofire, and Titov had the unnerving experience of pieces of it burning off as he descended through the atmosphere, similar to that experienced by Gagarin. Technical difficulties such as this were the subject of much debate and the process of continual reworking of systems for later flights. Yet for all the engineering work needed for those changes necessitated through post-flight analysis, Korolev was keen to push forward, and in September 1961 he proposed a triple-Vostok flight for November that year. One spacecraft would remain in space for three days, the other two for two or three days each, their missions overlapping by a full day.

Opposition to this plan was strong, and in October, believing they had seen enough spectacular flights, the Soviet government decided to cut Vostok flights in favour of boosting the launch of the spy satellite version known as Zenit. But then, yet again on a whim, direction changed once more. For some unknown reason, Lieutenant General Nikolai Kamanin, in charge of the cosmonaut group, began lobbying for a female cosmonaut, for which there had been no selection option with the first group. This ran into obstacles until Kamanin appealed direct to Khrushchev, who warmed to the idea.

To placate opponents, a decree issued by the Central Committee on 30 December authorised the appointment of 60 more cosmonauts, of which five were to be women. Then, quite suddenly, so as not to swamp the group, the selection of men was dropped and replaced by a call for 400 women candidates – of which five would be selected. This was *space real politik*, Soviet style!

While this was going on the first of Korolev's spy satellites was launched on 11 December, but a rocket failure prevented it from reaching orbit. When the spy satellite Object OD-2 had been integrated with the OD-1 manned programme and four different types for manned and photo-reconnaissance roles selected, the 2K became the spy satellite named Zenit, and then it got renamed Zenit-2. There was no Zenit-1.

While Korolev had got nowhere pushing for his triple-Vostok flight, he was given approval to go for a dual flight, and seven cosmonauts were short-listed. Suddenly, as though unaware of what had been going on in the United States, in January 1962 panic broke out amongst the Soviet hierarchy over US attempts to get John Glenn into orbit from Cape Canaveral aboard his Mercury spacecraft. Repeated postponements would push the first US manned orbit flight to 20 February. Korolev was ordered to get the dual flight up in March at whatever cost. But priorities were again imposed on getting

FAR LEFT In an unlikely parallel with Russia's R-7 derivatives that put Sputnik 1 and Vostok 1 in orbit, NASA's tiny Mercury spacecraft was tested on 15-minute suborbital flights using the Redstone rocket, an adaptation of the rocket that also sent the first US satellite into space. *(NASA)*

LEFT Not for more than ten months after the flight of Yuri Gagarin, and following the flight of Titov, did NASA manage to launch John Glenn on America's first orbital manned space mission on 20 February 1962. *(NASA)*

31

VOSTOK AND VOSKHOD

RIGHT Andrian Nikolayev prepares to ride the elevator up the side of the launch gantry to the Vostok 3 spacecraft, the first of two Russian spacecraft that would orbit the Earth together. *(Novosti)*

FAR RIGHT Pavel Popovich was launched in Vostok 4 one day after Nikolayev in a dual flight that was urged upon the Korolev team by Khrushchev, who was by now fully committed to using his cosmonauts as an effective propaganda tool. *(Novosti)*

a Zenit-2 into orbit, and that was finally accomplished on 26 April.

Delayed by work on the Zenit series, Korolev finally managed to get Vostok 3 into orbit on 11 August carrying Major Andrian Nikolayev, followed 23hr 32min later by the launch of Vostok 4 with Major Pavel Popovich aboard. While initially granting approval for these two flights to last three days and two days respectively, on 13 August it was decided to grant each an extra day in space, the two spacecraft landing within minutes of the other.

The dual flight was no mean feat, the challenge being welcomed by Korolev, now busy planning Moon missions for his Soyuz spacecraft. With two missions running concurrently, this would not be matched by the Americans until December 1965 when Gemini VI and VII were in space at the same time. But stretching mission duration to extract

RIGHT Popovich wears the spacesuit that was adapted from a military pressure suit used by MiG-21 pilots. The helmet was totally crafted for the Vostok missions. *(Novosti)*

FAR RIGHT Nikolayev, shortly after landing following almost four days in space aboard Vostok 3 in August 1962. *(Novosti)*

more publicity value was foolish and reckless. The temperature aboard Vostok 4 was falling dangerously low, although Nikolayev was all for staying up another day despite the cabin reading 50°F (10°C) with humidity up to 35%. But there were other signs: Popovich radioed that he could 'see thunderstorms', reiterating this pronouncement by saying that he was seeing 'a meteorological thunderstorm!'

Ever since Titov's flight, Russian physicians had been concerned about space sickness, and the leadership had not wanted to declare that Soviet cosmonauts were not up to the job. To this time the only American orbital flight had lasted three orbits and John Glenn had not suffered from space sickness. Time would show that only half of all space travellers develop this unpleasant side effect.

Aware that amateur radio enthusiasts around the world listened in to communications from space, Soviet cosmonauts had developed a set of code words for serious bouts of sickness and vomiting, indicating the imminence of such by declaring they were seeing 'thunderstorms'. Thus alerted, concerned about Popovich's state of health, and mindful of the falling temperatures in Vostok 4, it was immediately decided to revert to the original plan and bring them back down on the planned hour. Firing their retro-rockets within six minutes of each other, they landed a mere 124 miles (200km) apart. The irony was that Popovich had felt fine and really *had* seen thunderstorms.

ABOVE LEFT Nikolayev and Popovich confer over details of their dual mission in space, the first time two separate manned spacecraft would be in space together, and the only time for more than three years, until NASA placed Gemini VI and VII in orbit at the end of 1965. *(Novosti)*

ABOVE A highlight of the dual Vostok 3/4 mission was the live TV transmission to ground stations, where it was recorded and sent out to Russian stations. In this view, Popovich is in the top and bottom left during his flight while Nikolayev is seen at top right during a tele-recording. Popovich is at bottom right, seen prior to launch. *(Novosti)*

LEFT Achieving celebrity status known to only a few, Gagarin went on a world tour and served as ambassador for his country, vying with NASA astronauts who were required to do the same for the United States. *(Novosti)*

RIGHT Valeri Bykovsky was launched aboard Vostok 5 in June 1963 for the second dual flight mission of Russia's manned programme, his own flight lasting almost five days in space – a record to this day for a lone occupant. *(Novosti)*

BELOW Outside Moscow, a monitoring facility at the Ministry of Communications routes signals from the receiving stations tracking the flight of Bykovsky in Vostok 5. *(Novosti)*

The plan to send a woman into space envisaged a flight in 1963 and intensive training was undertaken to select prime and backup candidates from the four selectees of the five finalists. After much deliberation over whether it should be a dual flight of two women or a man and a woman, and after vacillation by the increasingly meddlesome bureaucracy that now surrounded these events, the commitment from the Central Committee on 21 March was to fly the mission that August. But when Korolev was told that all these delays had shifted the Vostok spacecraft beyond their shelf life, he successfully lobbied for a flight in June.

Vostok 5 was launched on 14 June 1963 with Valeri Bykovsky on board, followed by Vostok 6 with the world's first female cosmonaut, Valentina Tereshkova, two days later. The two were at their closest just a few hours into the flight when they were only 3 miles (5km) apart and some communication was possible between them.

Tereshkova suffered mild effects of space sickness and was unable to finish several experiments, encountering difficulty with mastering the manual orientation of the

BELOW Valentina Tereshkova was selected for the Vostok 6 flight, the final mission in Korolev's flagship programme, which brought immense propaganda value and genuinely turned world opinion regarding the capabilities of the Russian people. *(Novosti)*

RIGHT Valentina Ponomaryova was selected for a flight, but her demeanour was not considered appropriate for representing the USSR in the inevitable publicity that would follow the first flight of a woman in space. *(Novosti)*

FAR LEFT Irina Solovyova was Tereshkova's backup, but, although a candidate for a later space mission, never made a flight she had trained for so long to achieve. *(Novosti)*

LEFT Tereshkova was screened for her suitability to fulfil a demanding public relations role and was subject to a great deal of attention before and after her flight. *(Novosti)*

spacecraft, which she would have to do for re-entry if the automatic system failed. Much speculation was made in the West about her state of health, very little of which had any truth, and Tereshkova herself acknowledged the help she had received with instructions from the ground that enabled her to demonstrate mastery of the manual orientation for around 20 minutes while in space.

For his part, Bykovsky unstrapped himself and spent several orbits floating within the limited confines of the cabin, describing his experiences with enthusiasm and apparently not affected by the dreaded space sickness. But something he did share with Titov's mission: the equipment module again remained attached to the re-entry module and only released itself when heated in the atmosphere during descent, an unnerving threat to his safety brushed aside by the resolute cosmonaut.

BELOW LEFT Medical tests on Tereshkova were intensive and provided physicians with a suitable subject on which to test reactions on women in weightlessness. *(Novosti)*

BELOW Training for Tereshkova was no less strenuous than for the men and an essential part of preparing for a gruelling space flight. *(Novosti)*

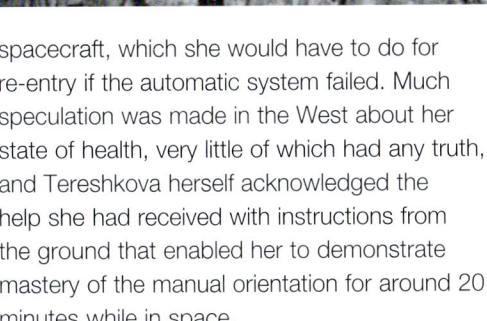

VOSTOK AND VOSKHOD

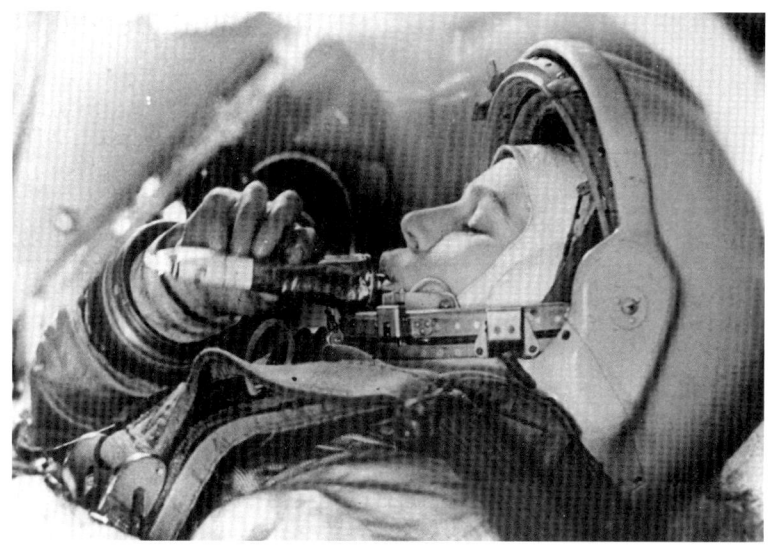

ABOVE Tereshkova spent more than 70 hours in space and sampled space food from tubes, slept, and monitored equipment, giving readings to the ground during the infrequent times the Vostok 6 spacecraft was within range. *(Novosti)*

Bykovsky's flight lasted 119hr 6min, still the longest solo space flight, while Tereshkova logged 70hr 50min. The two cosmonauts married in November 1963 and had a daughter, Yelena Andrianova on 8 June 1964, the first child born of two people who had been in space. Bykovsky would fly two Soyuz missions. The dual flight had been a tremendous propaganda coup and world tours added lustre to their reputation for heroism and for the technical accomplishment of the Soviet Union. Vostok flights were now becoming just that, with very little purpose in continued flights with a spacecraft of very limited capability.

With four spacecraft still available, Korolev wanted to use those potential flights as a stopgap to fill the void left by his technically delayed next-generation vehicle, Soyuz. He proposed a flight lasting up to 11 days carrying a dog to a highly elliptical orbit extending 620 miles (1,000km) above Earth to test radiation levels on living organisms. If that flight went well and there were no ill effects, he wanted to fly three additional Vostok missions of a similar duration, at the same altitude as the dog flight or in a regular low Earth orbit. Korolev wanted to make modifications to the 3KA to equip it with science experiments that the cosmonaut could work during the flights.

In the full knowledge that to support NASA's Moon goal the Americans were building a two-man Gemini vehicle capable of eclipsing Vostok's capabilities, it was important to Korolev to press ahead with Vostok's successor; but there seemed a general lack of will among those who made the final decisions about what Korolev and his team would be allowed to do. And there were disagreements too about just what should be done to counter

CENTRE A portent of the publicity to come, Premier Khrushchev talks on the telephone to Tereshkova, with President Brezhnev (right) and A.I. Mikoyan, the First Vice-Chairman of the Presidium of the Supreme Soviet. *(Novosti)*

LEFT Tereshkova and Bykovsky are welcomed at Moscow airport on 22 June 1963 after their historic dual flight, before driving to the city and a tumultuous welcome. *(Novosti)*

the American programme. Someone within the establishment of the political elite, a person clearly aware of not only the imminent Gemini flights but the three-man Apollo missions expected by 1966, encouraged Korolev in his desire for space 'firsts'.

Early in 1964 Korolev was asked to redesign the interior of Vostok to carry three cosmonauts, an idea which at first horrified him. All his inclinations were to press ahead with Soyuz using limited resources and talent now spread across several different programmes. In an attempt to show that there was a breakout from the one-man Vostok, the new programme would be called Voskhod (Russian for 'sunrise'). While upsetting the inertia essential to getting Soyuz under way, this interim digression would hold back progress with the next new vehicle.

The Military-Industrial Commission decided on 13 March 1964 that Korolev would adapt the four remaining Vostoks into multi-seat Voskhod spacecraft, creating the impression – purely for public relations purposes – that the USSR was already moving to next generation vehicles when in reality it was merely making use of stretched and overworked first-generation Vostoks. This version would be designated 3KV.

And then, just a few weeks later, a further order was issued requiring Korolev to carry out another adaptation by preparing for a spacewalk from an adapted Vostok carrying two cosmonauts, one of whom would go outside his spacecraft. This would be called Vykhod ('Exit'), designated 3KD. Supplements to this order authorised five new spacecraft: three Voskhods and two Vykhods. The formal authorisation for these missions was signed on 13 April 1964.

The move to be the first to have a spaceman leave his spacecraft was in direct response to NASA's published plans for the second manned Gemini flight, planned for early 1965, to conduct a spacewalk or EVA (extra-vehicular activity). In both of these mission directives, there was clearly a new and astute awareness of American plans, which evidence shows was not directly of concern to Khrushchev. Testimony from his son reveals that by this time the Soviet premier was getting tired of space projects and was unaware of the detail concerning NASA's plans.

Clearly, however, there was a group within the Soviet leadership which sought to influence the rocket engineers, proactively suggesting to them opportunities and possibilities far removed from a measured and prudent plan for extended space capabilities. Based more on propaganda than logic, it stretched Korolev's manpower and material resources to the limit.

The timetable for the 3KD was daunting, the flight being scheduled for September 1964. Under the guiding hand of Yevgeny A. Frolov, the Vostok would be entirely reworked. The first proposal, which horrified many, was the need to eliminate the ejection seat – thereby preventing any means of escape should the launch vehicle threaten to blow up – as well as to have the three crew members fly without suits. Neither could be avoided: there was simply insufficient room for three suited crew members. Moreover, and even more problematic, the crew would have to remain in the re-entry module all the way down to the ground.

ABOVE Broadcast on national radio and shown on TV transmissions later, Tereshkova begins her hardest journey in Moscow on 22 June 1963 – that of being a roving ambassador for the Soviet cause all over the world. *(Novosti)*

LEFT Established in 1933, the British Interplanetary Society hosts Valentina Tereshkova in London, here seen with President Dr Leslie R. Shepherd. *(David Baker)*

Also, there would be the need for a backup retro-rocket. On Vostok, if the rocket failed to fire the re-entry module occupied an orbit such as would naturally decay down through the atmosphere in any event within a few days. But with three people inside there would only be sufficient life support for one day, affording no margin for natural decay from orbit. The backup retro-propulsion would consist of three solid-propellant rocket motors delivering a thrust of 26,450kb (117.6kN) in a single two-second burst. Effective, considering it weighed only 500lb (227kg).

In the absence of an ejection seat, emergency escape from a launch vehicle running amok within the first 25–44 seconds of flight would be impossible. After that, the method would be the same as for Vostok, where the spacecraft itself would separate and land downrange. To effect a soft landing, the re-entry module would carry a triple parachute system for added braking, and a solid-propellant retro-rocket attached to the base of the parachute, which would fire only 4ft (1.2m) above the ground to decelerate the module from 18–22mph (8–10m/sec) to a mere 0.45mph (0.2m/sec). Ignition would be triggered by a line attached to the base of the re-entry module.

To say that this was a marginal mission is an understatement and reveals the extent to which the design teams were now in the grip of an ever-demanding political machine. On a one-person Vostok flight the cabin would lose 13.0 US gallons (50 litres) of oxygen per day through human consumption. On the Voskhod flight, the three men would require 47.5 US gallons (180 litres), thus limiting the flight to one day in duration. In total, Voskhod would have a length of approximately 16ft (5m), a diameter of 7.97ft (2.43m), and a mass of 12,530lb (5,682kg). The re-entry module itself would weigh 6,400lb (2,900kg).

Several features of the Voskhod spacecraft can be seen as precursor steps to the Soyuz spacecraft and as a bridge between Vostok and this later generation which had yet to be flown. One such example of how Voskhod served – unwittingly by its designers – as a bridge is in the spacecraft position and location indicator. Named Globus IMP, the initials stand for 'indication of position in flight' in Russian, the Voskhod device was developed from the Vostok instrument, which allowed the cosmonaut to see where his spacecraft was over the Earth. In effect, it showed the nadir of the orbit, the point directly beneath the spacecraft.

It was based on a rotating ball marked with continents and seas set within an electro-mechanical device. It was in essence a mechanical computer capable of computing complex functions and modulating electrical signals from other instruments on board. Although there would be considerable improvements, the device engineered for Voskhod remains the analogue against which all later derivatives were designed. The objectives met by its designers gave Globus the ability to provide the location of the spacecraft relative to a fixed frame of reference. It was an invaluable addition to the Vostok/Voskhod series because it would be the means by which the crew could conduct a manual backup burn of the retro-motor to bring the spacecraft out of orbit.

Due to the geographically isolated nature of the Soviet Union, most of each flight was spent out of contact with Mission Control. As well as providing visual indications that the spacecraft was performing as expected, Globus also allowed the pilot to know precisely where he or she was over the night portion of the Earth. At this time, representing almost 50% of a full orbit, the Vzor periscope and windows were useless for visual identification. The globe was also marked with the position of the tracking stations across the Soviet Union so that the cosmonaut would be prepared for a communication exchange.

Intriguingly, the paucity of contact opportunities forced development of the Globus device and for its adoption into the Soyuz programme. The Americans relied much more strongly on a network of ground stations providing much more frequent coverage and on maps carried aboard their spacecraft as well as computerised readouts on position. In Vostok, the instrument display system was set directly in front of the pilot, with the Globus panel directly in front above the Vzor screen. For the three-man Voskhod, however, the panel was rotated round to the side at 90° to the crew. The IMP panel itself was quite small, with a width of 9.8in (24.8cm), a height of 8.75in (22.2cm) and a depth of 5.75in (14.6cm).

The front of the panel was manufactured from machined aluminium alloy and the components were made of brass, with steel and aluminium where appropriate. The aluminium globe was covered with a thin covering of paper, in the tradition of most globes. The globe itself had a diameter of 5in (12.7cm) at a scale of 1:100 million with two degrees of freedom, giving it movement in rotation and inclination. A counter on the panel provided longitude and latitude marked in degrees, and an orbit counter logged the number of spacecraft revolutions of the Earth. This was divided into two white digits for orbits and red digits for fractions of an orbit. A status indicator light showed the point where the spacecraft would land.

The exposed portion of the globe was protected by a shaped plastic covering with a cross-shaped sight inscribed indicating the precise point of the spacecraft over the Earth; its nadir. In a manual setting, the cosmonaut could move the globe so that it would show where the spacecraft would land if the retro-rockets were to be fired at that moment, this also lighting an indicator. The latitude and longitude markers would automatically move to track these two positions. Before launch the Globus IMP would be set up with the coordinates of the spacecraft when it was expected to enter orbit.

After orbital insertion the ground tracking stations would provide an updated position that would be computed on the ground and read up to the cosmonaut. He would then reset the three correction parameters to effectively set the instrument in motion for the rest of the mission. He did this by adjusting the equatorial longitude and a given point in orbit, and in this way connected the flight sequencer to the solenoid actuator to send impulses which would convert the regular mechanical advance of the system to allow computations required for the instrument to move the globe and to set the other indicators in motion.

In addition, the variable resistor and electric blade contacts activated by a cam proportional to the rotation rate modulated signals from other instruments elsewhere on the spacecraft. This would deliver analogue signals providing a representation of the spacecraft's position over the Earth. All the while, the pilot would have the ability to reset the displays from information verbally transmitted from ground stations, re-calculated position plots generated in Mission Control and delivered back to the tracking stations.

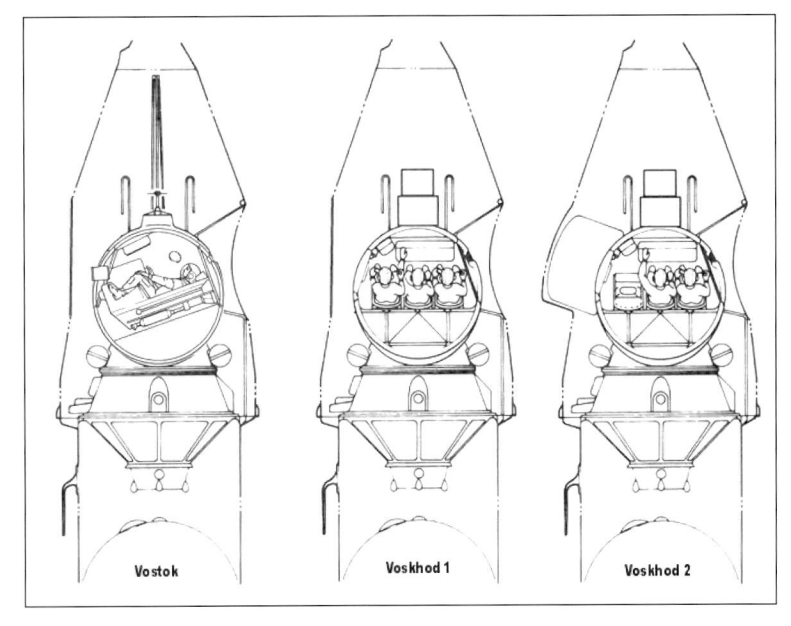

During the run up to the de-orbit burn, the cosmonaut would monitor the way the spacecraft was being orientated by the guidance and navigation system, and using the switch would move the IMP instrument forward 120° by activating an electric motor to relocate it to the predicted position of landing, progressively moving forward with the ground track. If the engine failed to fire, the cosmonaut could manually ignite the retro-rocket and the Globus would show him where, as a result of that burn, the spacecraft would land.

The Globus IMP was one of the finest demonstrations of how meticulous and fine-scale engineering could produce a mechanical computer, capable of operating in a vacuum should pressure in the cabin be lost, and from the incremental motion of the solenoid actuator provide such accurate information. With cardioid cam discs and cone-shaped ram cylinders the device is a horological delight, one which still impresses clockmakers today, and a device that represents the finest link between mechanical and electrical systems. It would serve as the baseline for a more sophisticated device in Soyuz.

The development of technology upgrades and improvements to the Vostok through its 3KV and 3KD derivates had relevance for the

ABOVE The evolution of Vostok to Voskhod masked attempts to promote the view that the multi-person adaptation was a completely new spacecraft, in part to show that Russia was ahead of America with their two-seat Gemini, then scheduled to fly in early 1965, and the three-seat Apollo due to fly in 1966 or 1967. The bulbous fairing on Voskhod 2 shrouds the collapsed airlock module. *(David Baker)*

Soyuz vehicle now solidly in development. But in mid-1964 the technical difficulties of producing a politically directed three-man flight was causing serious problems that delayed the planned flight date of August. Clearly, an automated precursor test in orbit was essential, especially as tests of the new parachute system with Titov's old Vostok capsule, retrieved from the secret OKB-1 museum, revealed serious problems. Nevertheless, disguised under the anonymity of Cosmos 47, 3KV-2 was launched on 6 October on a flight to demonstrate the new systems. It was a success and landed as planned after 24hr 18min, having circled the globe 17 times.

Selection of a crew for the manned flight was a point of contention, as Korolev had always wanted to have one of his engineers or scientists fly in one of his spacecraft. That had not been possible during the Vostok flights, but with three seats now available on one flight it was agreed that two of those seats could be filled by non-military personnel. The command seat would be retained for an Air Force officer. Several organisations got in on the act, and the selection process turned into a farce as different individuals tried to have their protégés accepted. The man from Korolev's organisation was Konstantin Feoktistov, but the choice for the third seat was hotly contested. In the end Korolev got his way and Boris Yegorov was appointed. Command of what would be publicly known as Voskhod 1 went to Vladimir Komarov.

Hailed as a completely new second-generation spacecraft, the launch of Voskhod 1 (spacecraft 3KV-3) took place on the morning of 12 October 1964 on an 11A57 rocket. The crew wore lightweight clothing and only communications hats incorporating headsets and microphones. They were the first cosmonauts to wear 'shirtsleeve' clothing, have no suits and no means of emergency escape. (No American astronaut would ever fly without a pressure suit as insurance against loss of pressure, and all vehicles except the Shuttle would have a launch escape system.)

After the customary telephone calls to senior officials of the Soviet government, and the traditional exchanges extolling the virtues of the mission and the enjoyment they were all having, the flight ended on schedule at an elapsed time of 24hr 17min 3sec. Again to worldwide acclaim, the mission attracted great publicity and triggered speculation about the design details of this 'new' spacecraft which had already achieved, in the eyes of the public, a three-person crew, several years ahead of the American three-man Apollo.

Within days of returning to Earth, Khrushchev had been deposed, his place on the Kremlin wall taken by Leonid Brezhnev to acknowledge the latest Soviet achievement. No longer would a single individual hold total power over the Party and the government. In a restructuring of space and missile organisations, the various bureaus involved in design and manufacture would be brought under the administrative wing of a newly formed Ministry of General Machine Building (Russian acronym MOM).

Pressure to get the 3KD spacecraft, equipped with seats for two cosmonauts, one of whom would leave his spacecraft on a spacewalk, was high. There was new technology here and a relatively unknown and unpredictable set of consequences with this risky procedure. The interior of the spacecraft could not be depressurised because there was insufficient life support function available, and concern was expressed about the ability to qualify a repressurisation system. So an airlock would be required into which the spacewalker would enter and from where he would exit once it had been depressurised.

This required a large appendage to the spacecraft and there was no room for a rigid structure, so a flexible, inflatable 'tunnel', with a hatch at the point of attachment to the side of the re-entry module and another at the other end, was the only way to achieve this feat. The deflated airlock would fold down like a concertina, gradually inflating in space as it was pressurised. Only then could the internal hatch be opened and the cosmonaut manoeuvre himself inside. But he would also need a spacesuit – another layer of technology.

All this was put together with remarkable speed and some testing. Like Voskhod 1, this Voskhod 2 mission had a political objective and required Korolev's team to beat the American Gemini spacecraft as it too approached the start of flight operations. Long after the collapse of the Soviet Union, rocket engineers and

RIGHT Pavel Belyayev (left) and Alexei Leonov were selected to carry out one of the most audacious missions of the entire Vostok/Voskhod programme – the first spacewalk. *(Novosti)*

managers testified to the incredible pressure applied to ensure this latest propaganda coup, enabled by a group who had no public recognition and whose very identity was kept a secret from the Russian people. They were even forbidden to speak to their wives of their association with these accomplishments.

Pavel Tsybin was responsible for the overall design of the 3KD, and S.I. Aleksandrov had the original idea of the inflatable airlock. When compressed, the airlock formed a circular shape with a diameter of 3.9ft (1.2m) and a depth of 2.3ft (0.7m), or 8.2ft (2.5m) fully extended. The internal diameter of the hatch was a mere 3.3ft (1m) while the diameter of the circular orifice to which the airlock was attached, and through which the cosmonaut would have to make his way in a fully inflated pressure suit, was only 2.1ft (0.65m).

Manufactured from rubber, the double skin consisted of 40 separate strips extending the full length of the airlock forming the circular tube. Air from two tanks would respectively inflate the airlock and then repressurise it after the EVA. Two additional air tanks were emergency supplies for the suited cosmonaut. In addition to a TV camera transmitting live pictures of the interior of Voskhod 2, two cameras were placed within the airlock and a third on a boom outside to record the historic event. Gay I. Severin was responsible for developing the airlock as well as the suit.

The pressure garment, named Berkut (Little Eagle) was fabricated from two layers designed to form an airtight seal when pressurised to 5.15lb/in^2 (266.5mm Hg). At this pressure the suit was almost rigid, and to improve flexibility, such as might be required in an emergency, a second setting would reduce pressure to 3lb/in^2 (155mm Hg), which would be dangerously low and run the risk of brain damage through oxygen starvation. The crew selected were at opposite ends of the selection criteria. Pavel Belyayev was a 39-year-old graduate of the top Russian military flight

LEFT Leonov was a vibrant 'star' of the first cosmonaut group and would have played a prominent role in Moon landings had they ever taken place. His selection for the Voskhod 2 flight was appropriate, given the demanding rigour of that effort. *(Novosti)*

BELOW Belyayev and Leonov were positioned side by side in the cramped quarters of what had been designed as a one-person spacecraft, each wearing a spacesuit. *(Novosti)*

ABOVE The rearrangement of equipment inside Voskhod was essential to providing sufficient space for the third seat in the case of Voskhod 1 and the airlock facility in Voskhod 2. *(David Baker)*

school, and Alexei Leonov was a 30-year-old combat pilot with a strong artistic talent and was the prankster of the group.

To test the new 3KD, Cosmos 57 was launched on 22 February 1965 as a full dress rehearsal of this planned one-day mission. All went well until a few hours after launch when the automatic re-entry system started, igniting the main engine. Unfortunately, being in an incorrect attitude for that manoeuvre at this phase of the mission, instead of re-entering the atmosphere Cosmos 57 ended up in the wrong orbit. Because its clock-activated destruct system was armed, it blew itself up before any of the technology related to the spacewalk could be tested. There had been a series of worrying failures with several of the new features added to the 3KD, and highly placed KGB officials suddenly turned up unannounced and placed time-consuming security restrictions on the pace of the work.

Of worrying concern was the impact that the airlock adapter ring placed on the outside of the descent module would have on re-entry. Because that had not been demonstrated on Cosmos 57, a routinely scheduled Zenit-4 spy satellite – these now being launched at frequent intervals – was fitted with a mock ring and launched as Cosmos 59 on 7 March 1965. Recovered eight days later, with rolls of film from its primary function, the ring was found intact and not to have affected re-entry, although there were indications that it had set up a spin rate of almost 100°/sec – very uncomfortable for a crew, but survivable.

In America, preparations for the launch of the first manned Gemini spacecraft were under way, and in Russia Korolev cautioned the Voskhod's crew against 'thoughtless heroics'. Suffering from a severe pulmonary infection, Korolev was not well and there were concerns for his life. But the need to beat the Americans was now a driven imperative, urged on by political considerations but felt deeply as a matter of pride by the nameless thousands who had fulfilled the dreams of their chief designer. In common with dedicated scientists and engineers the world over, they were for the most part content to see their greatest hopes fulfilled and did not seek personal glory.

Voskhod 2 was launched on 18 March from a snow-bound launch complex, spacecraft 3KD-4 being placed into a nominal trajectory despite dire warnings of near-failure to several systems during ascent. The sole purpose of this mission was to beat the Americans to a spacewalk, and preparations began almost immediately. Stocky yet taut and muscular, Leonov was to endure physical stresses during the next few hours unparalleled in any other mission. His training had prepared him for it. On one demanding test of physical and psychological endurance, he had been placed in an isolation chamber totally devoid of any sensory connection with the outside world for one month. Taken out and put straight into a jet fighter, which proceeded to throw him around the sky with high aerobatic g-forces, he was required to eject himself and land by parachute!

Coming back across the Mediterranean toward the end of the first orbit Leonov was already in his suit, the airlock inflated and the inner hatch open. Impatiently waiting for permission to depart, he squeezed into the flexible airlock, closed the inner hatch and opened the hatch in front to the vacuum of space. Flying over the Black Sea he removed a cover on the exterior camera, raised the boom to which it was attached, and pushed himself outside. Elated by the view below him, the Earth slowly rotating before his eyes, Leonov let go his grip and floated free secured only by a 17.5ft (5.35m) tether. As he watched the brightly coloured Earth below he drifted across the southern part of the USSR and on to his second orbit.

Inflated to the point where the suit had extended beyond his hands and feet, Leonov found it impossible to get back into the now limp unpressurised airlock section. His suit was simply too bloated for the size of the hatch ring. Deflating his suit to the lowest setting (3lb/in²; 155mm Hg), he finally found he could use his hands to get a grip on the outer hatch and work his way inside the airlock. Outside for little more than 12 minutes, he was struggling now with over-exertion, his pulse at 143 beats/min, high respiration rate and a body temperature of 100°F (38°C).

With the outer hatch closed, the airlock section could be repressurised and the inner hatch opened to the re-entry module. His EVA time, recorded as the duration in which the outer hatch was open, was 23min 41sec. Close to being overwhelmed by body heat and exhaustion, contrary to the planned procedure Leonov had opened his suit faceplate inside the pressurising airlock and spent several minutes resting.

As the flight progressed through its planned full day in orbit, the crew discovered that the hatch had not sealed correctly and that the air was slowly bleeding away. Overworked by the need for replenishment, the automatic system pumped higher quantities of oxygen into the cabin, raising the level to 45% of the atmosphere. Russian engineers were terrified of a flash fire caused by an electrical spark and both cosmonauts worked feverishly to get control of the replenishment system. They finally managed to get it down to a safer level, but the fate of Bondarenko haunted them.

When time came for re-entry, the engine failed to fire and the crew had to wait another orbit to accomplish this manually. As the spacecraft had been redesigned inside to accommodate the two seats, the Vzor optical alignment device was displaced 90° to its original position in the one-person Vostok. The space-suited Belyayev had to position his body transversely across the two seats while Leonov found a position half under his left seat holding

LEFT The inflated airlock module on Voskhod 2 enabled Leonov to exit the spacecraft without depressurising the cabin. Cameras (1) allowed images and film to be taken of the historic event. *(David Baker)*

ABOVE **Leonov emerges from the depressurised airlock, the top hatch seen to one side.** *(Novosti)*

BELOW **Leonov emerges fully, the hatch folded back to the right side.** *(Novosti)*

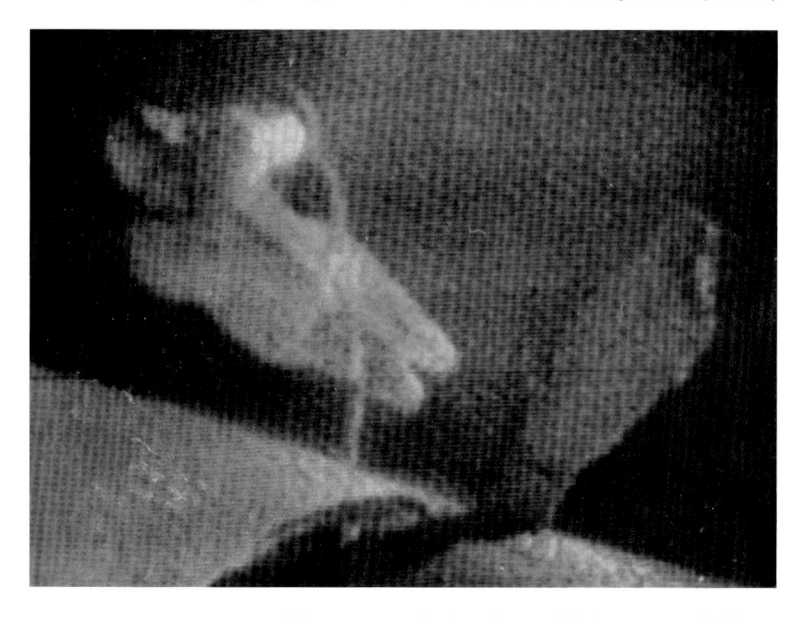

his commander in place while he got a sighting on the appropriate time and place to fire the retro-engine. It took Belyayev 46 seconds to return to his seat and push the ignition button, but Leonov did not make it to his seat in time. It was vital that both men were strapped in their seats at retrofire to get a fully balanced weight distribution for attitude hold and to prevent the spacecraft spinning out of control.

Already nearly a minute late hitting the ignition button, the spacecraft hurtled into a steeper re-entry path than normal. Then the equipment section failed to release, hung up by the straps that should have broken free.

This combination of unfortunate events brought Voskhod 2 down at a steeper angle, imposing 10g on the two cosmonauts. Landing farther down-track than expected, 26hr 2min 17sec after launch and in deep snow in a densely forested area, it was four hours before anyone knew if they were alive. Had the re-entry module brushed the treetops the braking rocket would have fired briefly high off the ground, causing the sphere to drop like a discarded stone. When the recovery team arrived next day on skis after a freezing night, they found Belyayev and Leonov gathering wood for a fire.

As related in the following section, by early 1965 the Soyuz spacecraft was well under way, as were schemes to fly cosmonauts around the Moon and – a more ambitious venture – to land on its surface, which had only been given full authority to proceed the previous year. Plans for further Voskhod flights to fill the gap before Soyuz flew were quite ambitious.

In July or August, 3KV-5 would fly two dogs on a 'Cosmos' mission high inclination orbit for 15–30 days, in September or October Voskhod 3 would carry a pilot and a scientist on a 15-day flight with several experiments, Voskhod 4 would follow in March or April 1966 on a mission lasting 15–18 days, and Voskhod 5 would conduct another EVA on a mission lasting up to five days, repeated by Voskhod 6 before the end of 1966.

These missions were firmly planned to challenge the potential success with NASA's two-man Gemini flights, where spacewalking, 14-day missions and rendezvous and docking were to be demonstrated in a series of flights running at approximately two-month intervals. The Russians too were developing a backpack that a cosmonaut could wear for manoeuvring around during a spacewalk, another plan developed for the much-publicised Gemini programme. However, continuous delays brought about by technical challenges, and changes to flight objectives to keep pace with the gathering momentum of the American programme, doomed all these ambitious Russian expectations.

On 16 January 1966, with the Voskhod programme stumbling toward a series of missions that now aimed to complement the string of Soyuz flights planned for that year, the official Soviet news agency, Tass, announced in grim tones the death of Korolev.

He had succumbed to a massive tumour, the magnitude of which was only discovered when he lay on the operating table in a Moscow hospital. Suddenly, from total obscurity, the name of Sergei Korolev was trumpeted from the Kremlin walls, across the USSR and out to the wider world. With an outpouring of emotion and sustained eulogies, the architect of Russia's extraordinary space programme, the victim for several decades of a Soviet system he spent his life serving, was no more.

Korolev was replaced at the helm of OKB-1 by his long-serving deputy, Vasili Mishin, to whose responsibility fell the management of a wide range of space projects, not least the continuation of the Voskhod programme and the introduction of the second-generation Soyuz into operational service. In preparation for the flight of Voskhod 3, planned for 10–20 March 1966, two dogs – Veterok and Ugolek – were launched in 3KV-5 on 22 February.

Designated Cosmos 110, the spacecraft was placed in an elliptical orbit with a high point (apogee) of 561 miles (904km) for a flight that lasted almost 22 days. The high apogee exposed the dogs to the lower levels of the Earth's magnetic radiation belts to determine if they suffered any ill effects. One dog had been fed anti-radiation pills, the other was the control and both survived their ordeal, albeit in an advanced state of deterioration. The dogs experienced all the unpleasant side effects of long-duration space flight, including calcium loss from their bones, dehydration, loss of muscle and difficulty in standing and walking – symptoms which lasted for nearly two weeks after their return.

Cosmos 110 had been precursor to the 18-day flight of Vostok 3, but that mission never flew and 3KV-6 remained grounded. Grave concerns about the ability of the now aged design to fly long-duration missions in safety arose from test failures to the life support systems being designed for these flights. Results from Cosmos 110 revealed a poor performance from this system and an increasingly bad cabin atmosphere, which would have brought a decision to terminate the flight much earlier than it was had humans been on board. As it was it was felt wise to bring the dogs down three days short of the planned 25 days.

By June 1966 it was clear that the Voskhod programme had run its course and, with the Soyuz programme approaching its initial flights, the legacy of Vostok would have to be seen in that vehicle and not the sustained use of an ageing Voskhod originally intended as a propaganda stopgap against American achievements. It is an ironic footnote to the Voskhod story that the fulfilment of Korolev's ambitious set of mission objectives for his multi-seat spacecraft – extensive spacewalking, long-duration flight and repeated two-seat operations – were fulfilled in space by NASA's Gemini programme and not by Voskhod.

LEFT A sequence from the film camera that recorded the world's first spacewalk, which lasted a little longer than planned due to the difficulty Leonov encountered getting back in. *(Novosti)*

ABOVE Cosmos 110 carried the dogs Veterok and Ugolek into space on 10 March 1966 aboard the Vostok spacecraft 3KV-5, an event commemorated on this stamp, and the last flight of living things aboard a Vostok. *(David Baker)*

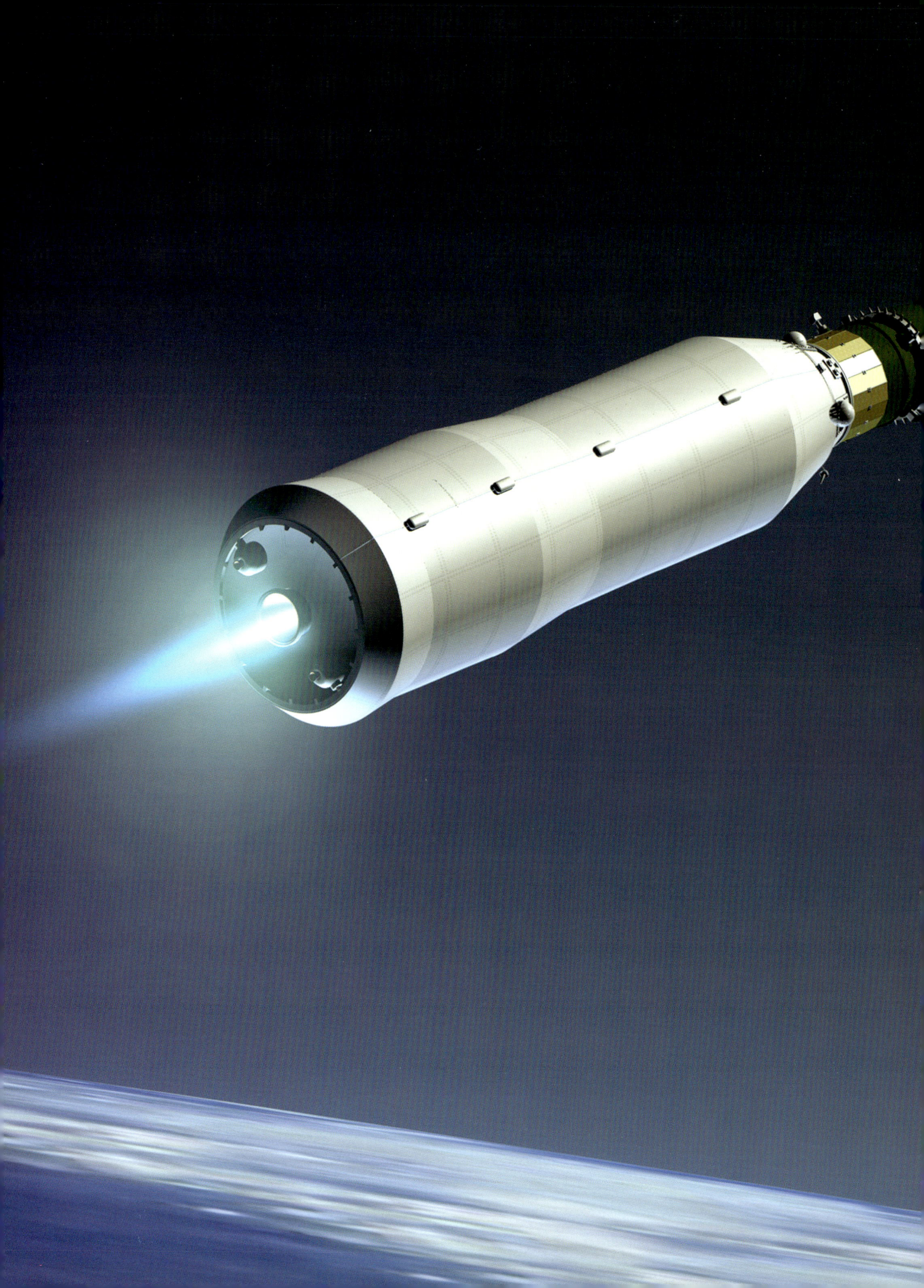

Chapter Three

The 7K-OK Soyuz

Long before the end of the Vostok/Voskhod programme, Russian space engineers were hard at work designing the next generation of human space vehicles. But soon there was to be a fierce debate not only about the next goals for Soviet manned space flight but as to who should be selected to get the prime job.

OPPOSITE With the shroud covering the LOK jettisoned toward the end of powered flight into a preliminary Earth orbit, the Block G rocket stage ignites to propel the remaining elements toward the Moon. The Block C stage and the LK lander above it remain shrouded until the Block G is jettisoned. *(David Baker)*

ABOVE During the period 1961–63 Soviet Premier Nikita Khrushchev and US President John F. Kennedy engaged in a battle for global influence that fuelled the space programme into a technological race for prestige and influence abroad. *(White House)*

RIGHT NASA's Gemini programme had an early success when, on 3 June 1965, astronaut Edward White II conducted a spacewalk from his Gemini IV spacecraft to equal the feat performed by Alexei Leonov in Voskhod 2 almost three months before. *(NASA)*

For Korolev, whose design team were the sole practitioners in the field, any competition at this stage was unthinkable. But these confrontations would come later. At OKB-1, two years before Yuri Gagarin would become the first man in space, the future was already on the drawing boards.

No sooner had the Soviet government given its official approval to the start of a human space flight programme in January 1959 than Korolev recognised that a successor to the simple Vostok would be needed to maintain momentum. For a year already, his OKB-1 design team had been examining a successor. The assumption was that flights around the Moon would be followed by a space station in earth orbit and manned landings on the lunar surface. All this was a precursor to manned flights to the planets, especially to Mars, which Korolev pinned his hopes on reaching in the foreseeable future.

The lionised Russian space prophet Konstantin Tsiolkovsky had laid down the road-map for space flight before the 1914–18 war, and for Korolev this made sense. He saw circumlunar flight as relatively simple, calling for a single spacecraft to be propelled to such a high orbit that it would be captured by the Moon's gravity and in a slingshot move, thrown around the far side and back to Earth. A space station would need to be big, as would a spacecraft sent to Mars, and the ability for one spacecraft to catch up with another and dock to it was, in Korolev's mind, the fundamental enabling capability of future space flight.

First would come the circumlunar flight to achieve worldwide acclaim and to counter critics in the Kremlin, who claimed that this was a diversion of national assets that would bleed defence money into showcase spectaculars of little real value. But Khrushchev was sold on the propaganda value, seeing in ever more ambitious space 'firsts' a tool every bit as powerful as expansive military might. In this he was ahead of the United States, where political judgement was ambivalent about space and largely unwilling to expand upon its slow and cautionary progress with the tiny Mercury capsule and other low-key space projects. Until, that is, Yuri Gagarin demonstrated the appeal for a world agog at this fantasy-turned-fact.

Meanwhile, in April 1959, a spacecraft design concept emerged for achieving orbital rendezvous and docking. Known as the Sever (North), it was a modest step beyond Vostok, capable of carrying three cosmonauts in spacesuits. Another concept emerged from the drawing board of Pavel V. Tsybin, a much more ambitious design capable of carrying seven crew members, which ousted a spaceplane concept such as had been studied several years earlier. The larger spacecraft was quickly discarded as being too ambitious. From these studies came a competitive design process in two separate departments of Korolev's OKB-1 which for the next year bounced ideas back and forth.

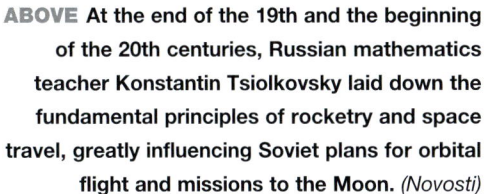

ABOVE At the end of the 19th and the beginning of the 20th centuries, Russian mathematics teacher Konstantin Tsiolkovsky laid down the fundamental principles of rocketry and space travel, greatly influencing Soviet plans for orbital flight and missions to the Moon. *(Novosti)*

RIGHT Tsiolkovsky's garden is preserved today, a place where he used to deliberate on the great challenges facing humans in the conquest of space. *(David Baker)*

BELOW Kaluga now boasts a museum containing exhibits from Tsiolkovsky's life and much memorabilia associated with his prophetic writings. *(David Baker)*

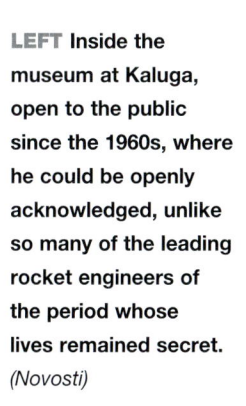

ABOVE Tsiolkovsky was brought up in the small town of Kaluga and lived in this house, where he did most of his writing and made models to demonstrate the design and layout of a basic rocket. *(Novosti)*

LEFT Inside the museum at Kaluga, open to the public since the 1960s, where he could be openly acknowledged, unlike so many of the leading rocket engineers of the period whose lives remained secret. *(Novosti)*

RIGHT Adopted by the Communist Party as an idealistic representation of socialist ideals, Tsiolkovsky's monument was erected in Moscow at the entrance to another museum exhibiting spacecraft and rockets. *(David Baker)*

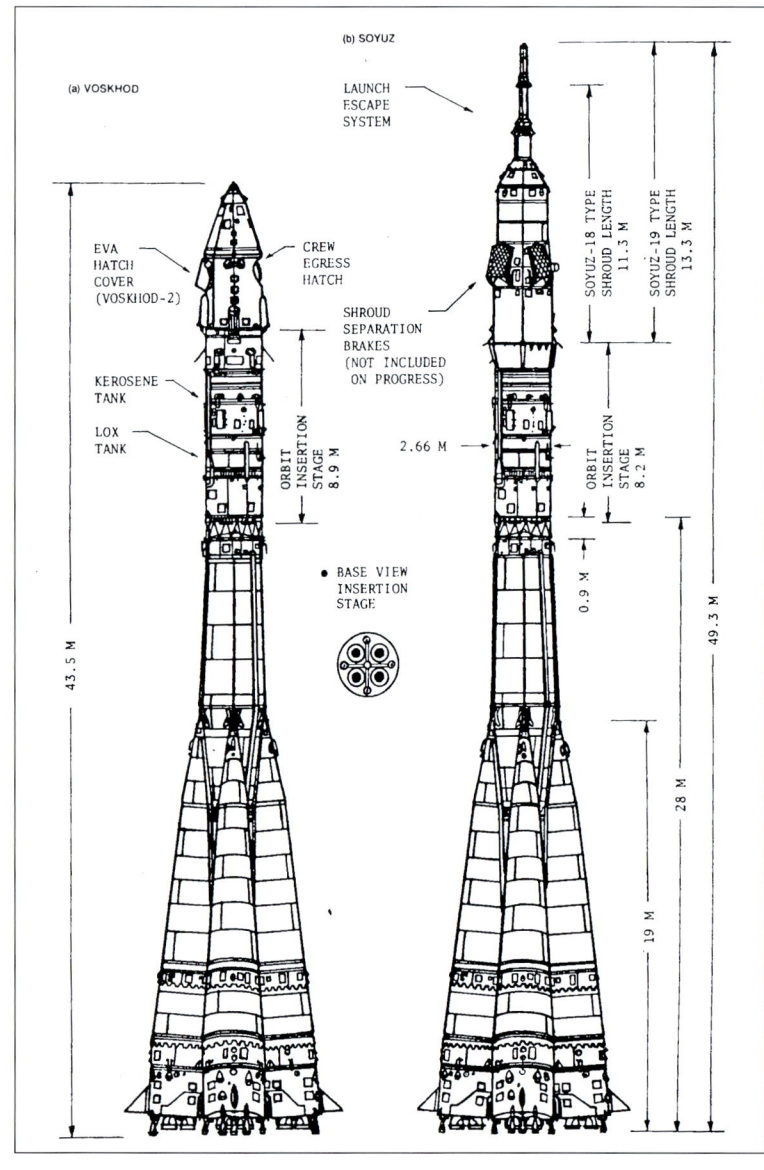

In 1960 and 1961 the design of the Sever began to take shape, a spacecraft designed through the several different analyses conducted by aerodynamics laboratories and research institutes. Unlike the spherical Vostok/Voskhod series, they favoured a re-entry module shaped like a car headlight – the base being approximately the diameter of the height. With this design the spacecraft would stabilise itself during the long plunge through the Earth's atmosphere and provide a more secure and safer ride to the ground because as it penetrated deeper the lift ratio would increase, allowing the trajectory to be less severe. Attached to the descent module, an equipment section would provide power, communications equipment and propulsion units until shortly before re-entry, when it would be discarded.

By 1962 Russia and America were both flying astronauts into space, and NASA had been given the task of racing to the Moon by the end of the decade. While they thought a flight around the Moon and back to be a relatively easy task, the Russians believed that the landing envisaged for Apollo was simply too ambitious to have any hope of success by the end of the decade. The result was that for the next two years, Soviet programmes would mature along a separate line of development, only joining the race to the lunar surface in 1964.

Korolev's train

Since early 1962 Korolev had been working on a circumlunar mission that he envisaged would involve a train of five modules launched separately. The spacecraft design selected for this mission was designated 1L and was a hybrid adaptation of the Sever ship incorporating a 'habitability' module where the crew could relax during the six-day flight out to the Moon and back. A lengthy analysis of possible configurations decided upon a spacecraft with four sections arranged in tandem: at the front the habitation module,

LEFT Development of the R-7 took a step further from the launch vehicle used for Voskhod (left) with the development of the rocket that would lift Soyuz. *(David Baker)*

a second module for the crew to occupy during launch and landing, a third module for instrumentation and propulsion and a fourth module for rendezvous equipment.

Korolev knew that he would have to find a military application for his proposed Sever to justify development of the spacecraft and have a vehicle capable of circumlunar flight. So in the proposal he pointed to Sever's potential use for satellite interception and destruction, for launching sophisticated military communication satellites to geostationary orbit and for supplying a military space station in Earth orbit. The circumlunar flight would require additional rocket stages to push 1L to the Moon. These he linked to the need for additional rocket stages for placing satellites in geostationary orbit, this justifying their development too.

Korolev's plan was complex and involved a flotilla of vehicles and rocket stages he called Soyuz (Union). All schemes would have to be capable of flying on top of the adapted versions of the R-7 with more powerful upper stages, and this limited their size. The Soyuz train would involve the new manned spacecraft, the IL and three small rocket stages aligned in tandem for powering the assembly to the Moon. It would also involve a modified Vostok 7 spacecraft, uniquely adapted for a specific role as Voskhod would be.

Launched first, a new and much modified Vostok 7 would have been fitted with a propulsion module for changing orbit, allowing it to conduct orbital rendezvous and docking. At an appropriate time, at one-day intervals, successive launches would each deliver a fully fuelled rocket stage into orbit weighing 10,600lb (4,800kg) apiece, each of which would automatically dock to the other in tandem. Still attached to the base of the first rocket stage, Vostok 7 would be the coordinating assembly ship. Finally, the 1L ship would lift its three-person crew to orbit and dock on to the opposite end of the train to Vostok 7, leaving the latter and its pilot to return to Earth, its job done. In this configuration the three rocket stages, called blocks, would fire in sequence to boost the 1L spacecraft around the Moon and back.

The technical prospectus was issued by OKB-1 on 16 April 1962 and the Soviet leadership signed up to the scheme but nothing

ABOVE By the end of 1965 America had achieved all the primary objectives of its Gemini programme except docking with another vehicle in space, including the first rendezvous between two objects launched from Earth, which was accomplished when Gemini VI and VII met up in space during December 1965. *(NASA)*

BELOW A Gemini spacecraft manoeuvres towards an Agena rocket stage with which it will shortly dock. Many such dockings took place on four Gemini missions during 1966, a feat that would not be equalled by the Russians for three years but one which would be crucial to their Moon landing plans. *(NASA)*

happened. No resources were apportioned from the central bureaucracy and Korolev came up against a wall of silence and indifference. Enter Vladimir Chelomei, the fast-rising star of the emerging space programme, as a serious and highly political opponent of Korolev and critical of the scheme.

Very quickly, is seems, the 7K/1L proposal was seen as too complicated and too reliant on an ageing Vostok spacecraft design for a mission whose success would hinge on four rendezvous and docking activities in rapid succession. Within months the project was abandoned and Korolev had taken the concept of the much modified Vostok and incorporated its identification number into a new design – the 7K, which now was known as Soyuz, the name previously adopted for the rocket train to Moon orbit. This new ship would merge the best elements of both Sever and 1L in a spacecraft that could conduct a multiplicity of roles.

As refined, the configuration of the Soyuz spacecraft was established in this late 1962 rework. The descent module would carry the two- or three-person crew, an orbital module would provide habitation and be attached to the front and an instrument module would be attached below to house all the support systems and the propulsion units, plus solar arrays providing electrical power essential for long flights.

At one point the orbital module had been placed beneath the descent module because engineers wanted a launch escape system that would carry the crew away to safety in the event of a malfunction during ascent. But that would have meant having a hatch in the heat shield at the base of the DM, and Korolev was unable to accept the risk in penetrating a critical part of the structure, essential for survival during re-entry. In the end, the complete spacecraft would be lifted free in the event of an abort.

Korolev signed off on the 7K/Soyuz design on 7 March 1963 and it was submitted to the political leadership on 20 March. On 10 May he submitted a further plan, for using this new spacecraft to achieve a circumlunar flight using the basic vehicle without support of a Vostok derivative, now abandoned. This time, in hearings before a technical council, Korolev got the grudging approval of Chelomei and on

ABOVE Represented here by a much later launch of the Proton rocket, Chelomei's UR-500K would challenge Korolev's plans for circumlunar flight, and stall plans for a direct manned flight around the Moon. *(David Baker)*

RIGHT The general configuration of what would become the Soyuz spacecraft was defined by the early 1960s, in parallel with Vostok in what Korolev considered was a precursor to an operational vehicle. *(David Baker)*

the basis of this proclaimed the first test flight in 1964. But there was a reason for Chelomei's support. With the need for additional rocket stages to propel Soyuz around the Moon, there would be additional opportunities for contracts.

Korolev's refined circumlunar mission would now involve first the launch of a rocket stage to low Earth orbit. Designated 9K, this stage would be topped up by at least four 11K tanker stages launched in succession, each equipped with automatic rendezvous and docking equipment and with pipes interconnecting for transferring the 9,160lb (4,155kg) of propellants to the 9K. In total, the 9K would contain approximately 45,000lb (99,225kg) of propellants. Then the 7K/Soyuz and its crew would be launched and docked to the front of the 9K. The 9K stage would ignite its 9,900lb (44.135kN) thrust rocket motor propelling itself and the 7K/Soyuz to the Moon. Once the Soyuz separated, the main propulsion engine on the 7K would be used for trajectory correction manoeuvres on the way to and from the Moon.

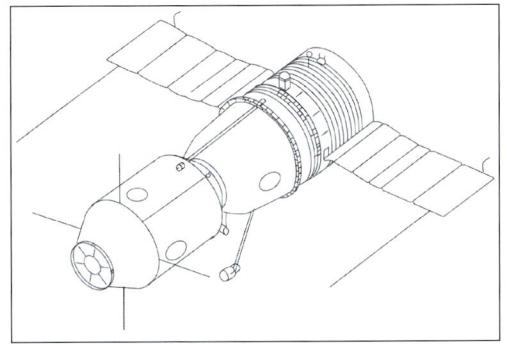

ABOVE To cushion the impact of landing, a series of small retro-rockets were built into the underside of the descent module which was exposed when the base heat shield was jettisoned and was activated very close to the ground. *(Energia)*

LEFT Unlike Vostok, Soyuz crew members would have to remain inside their spacecraft for a landing, and with the additional weight of the descent module there would be a risk of injury. The entire module had to be redesigned to accommodate a shock-absorbing system within the seat structure. *(NASA)*

BELOW The basic premise behind the design of what was referred to as Soyuz-A was the need to separate the module that re-entered the atmosphere with a work station which would be jettisoned after retrofire. The equipment section in the back would contain all the support systems for orbital activity. The early version of the work module had a cylindrical centre section with end closures, a shape that would evolve into an ellipsoid. *(David Baker)*

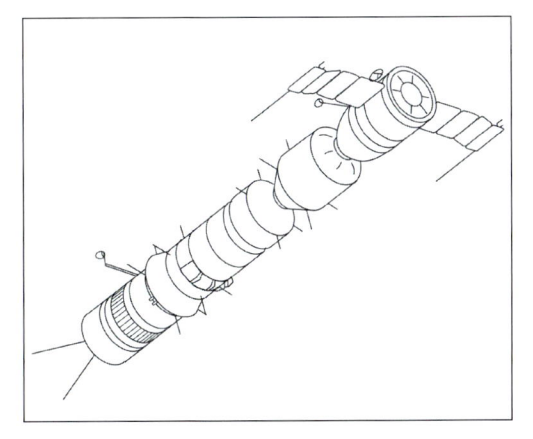

LEFT An early concept for circumlunar flight involved three separate elements comprising a Soyuz-A spacecraft (right) docked to a propulsion unit (centre) and a tanker module (left). *(David Baker)*

THE 7K-OK SOYUZ

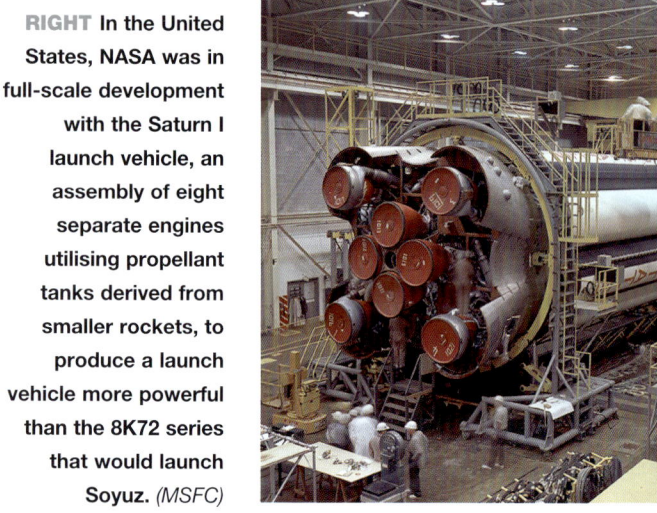

RIGHT In the United States, NASA was in full-scale development with the Saturn I launch vehicle, an assembly of eight separate engines utilising propellant tanks derived from smaller rockets, to produce a launch vehicle more powerful than the 8K72 series that would launch Soyuz. *(MSFC)*

BELOW Saturn I flights began in October 1961 and would continue with orbital tests of the S-IV cryogenic upper stage, super-cold liquid hydrogen/oxygen propellant which would be adopted for the second and third stages of the Saturn V, NASA's Moon rocket. *(NASA)*

ABOVE In America, NASA pinned its hopes for a Moon landing by the end of the 1960s on a massive new facility built at Cape Canaveral. Known as the Kennedy Space Center, in honour of the late President, it had the Vehicle Assembly Building as its centrepiece with a five-mile (7km) road leading to two raised launch pads. The Saturn V would be carried to its selected pad on a mobile platform on which it had been erected and transported by a crawler vehicle beneath. *(KSC)*

In no small measure, the Russian space programme was surging to a peak of work that was beyond the infrastructure available. An industry accelerated by political desire for propaganda and by a group of highly motivated, driven idealists intent on racing to the Moon and the planets was incapable of adequately supporting the increasing number of major projects. It was no surprise that the work on the 9K and 11K elements were to be passed across to other design bureaus, leaving Korolev to concentrate on developing the 7K/Soyuz.

As recently as 24 September 1962, the Soviet Council of Ministers had sanctioned development of a giant super-rocket designated N1 capable of placing 75 tons in Earth orbit, more than ten times the capability of the existing rockets used to launch Vostok and Voskhod. With Korolev proclaiming the first flight of the N1 in 1965, bigger versions were proposed with liquid hydrogen/liquid oxygen upper stages and capable of placing almost 100 tons in Earth orbit. This, and encroaching work on the Soyuz spacecraft, plus the planning for advanced Voskhod versions of Vostok, placed almost unbearable strain on existing manpower and resources.

While work was under way, and in the wake of approval from the Council of Ministers, Korolev again ran into obstacles. Although a political agreement had been reached to go ahead, the financial side of the programme was running expensively over budget, and while the Soyuz was being funded under the Air Force no one was taking responsibility for the costs of the circumlunar segment. This opened a path once more for Chelomei, who had been developing his own rocket, the UR-500K, soon to be known as the Proton launcher, after the name of the first satellite it would carry into space.

While the R-7 had been developed as a ballistic missile and then adapted as a satellite launcher, the UR-500K was specifically designed as a multi-purpose heavy-lift satellite launcher and ballistic missile with a capability of carrying a massive nuclear warhead. Chelomei had been in charge of the OKB-52 design bureau engaged in development of cruise missiles for the Navy and quickly reacted to the decision by Khrushchev in the late 1950s to switch arms production from bombers to missiles. In 1960 he began formulating ideas about satellite launchers through a project

BELOW The various stages of the N1 were massive, with 30 engines in the first stage, eight in the second stage and four in the third stage. The payload section comprising Soyuz-derivative LOK and Lunar Lander LK spacecraft together with relevant propulsion stages measured 141.7ft (43.2m) tall on top of the three-stage N1. *(David Baker)*

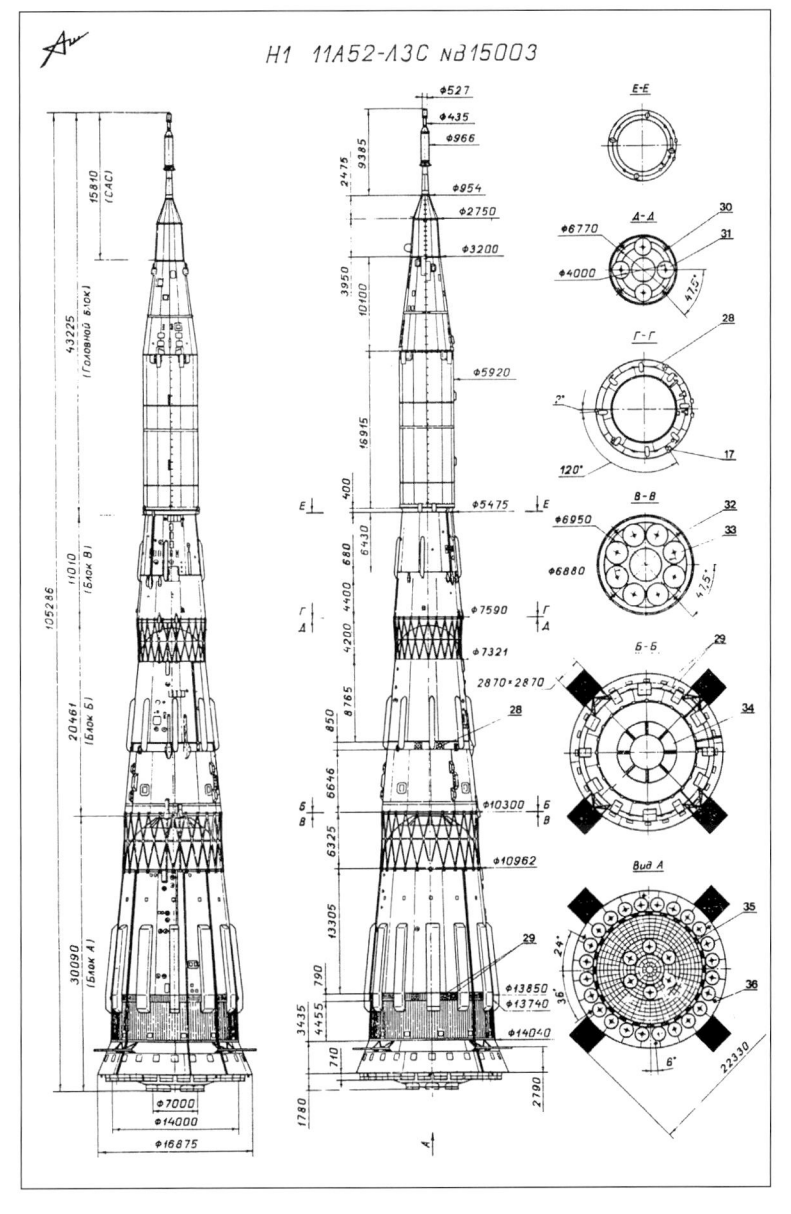

BELOW The massive N1 rocket compared in size to the Saturn V, both equally tall, but the N1 had a greater thrust but less payload capability than the NASA rocket because the latter used high-energy upper-stage propellants. *(David Baker)*

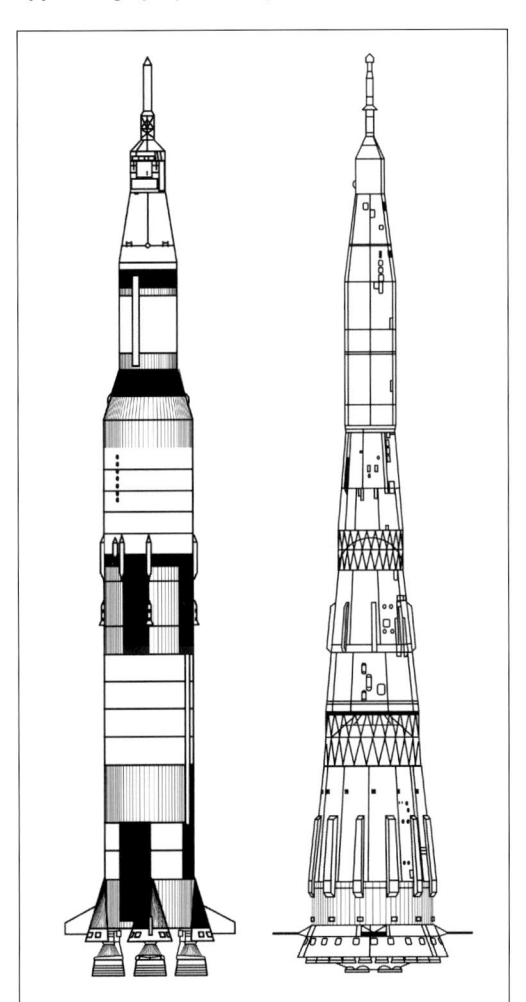

RIGHT The Soyuz LOK Moon-orbiting mother ship was similar in length to the Apollo command and service modules, but unlike the US system the Russian LOK required a separate rocket stage. *(David Baker)*

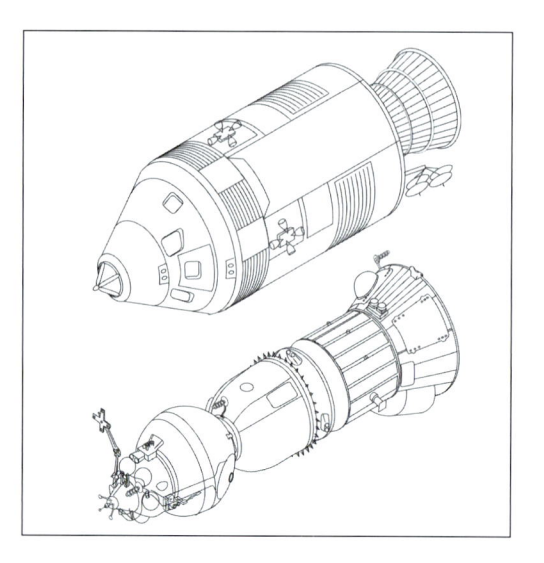

known as the A-300 and got Khrushchev's approval to develop it, which is not overly surprising since Khrushchev's son worked for the rocket designer.

This led to the A-600, which was nearly cancelled when the Russian military rebelled against Khrushchev's over-indulgence in space projects and a decree was obtained limiting these activities in 1961. Eventually, officially designated the 8K82, the UR-500K emerged from this cycle with approval to proceed granted in February 1962. The potential of this rocket, with its ability to launch more than 44,000lb (almost 20,000kg) into orbit, was a significant increase on the payload capacity of the 10,400lb (4,700kg) capability of the 8K72K used for Vostok flights.

Chelomei's UR-500K was the perfect weapon with which to challenge Korolev and his giant N1, which even by late 1962 was running into serious development problems. Promising to achieve a circumlunar flight long before Korolev with his 7K/Soyuz, Chelomei directly challenged the previously agreed plan and set out what he promised would be a faster and less expensive route to flying around the Moon. But he was not alone in promising a multi-functional purpose to what was ostensibly a space project. Korolev too understood that the military, incensed by the preoccupation with space projects, would only give their approval if they were a significant asset for them.

In presenting the revised 7K/Soyuz scheme in May 1963, Khrushchev had proposed two military versions of his spacecraft. One, the Soyuz-P, would be developed as a space-based interceptor, using its main propulsion system to hunt down and destroy selected satellites at will up to an altitude of 3,730 miles (6,000km). The second version, Soyuz-R, would perform a reconnaissance function as well as carrying out a variety of military activities in space. The Perekhvatchik interceptor would have used the 9K/11K rocket stages

LEFT The separate stages to N1 each contained spherical liquid oxygen and kerosene tanks in an arrangement that emphasised rigidity, and without the sophisticated engineering technologies used on NASA's Saturn V. *(David Baker)*

for achieving high altitude orbits in its military role. This project was assigned to Department 3 under Dmitri Kozlov, as was the Razvedki reconnaissance version, which would have had a dedicated section for surveillance equipment.

Korolev also proposed a small space station that would be served by Soyuz, thus anticipating the ultimate role of that spacecraft several years later. But this was a desperate attempt to gain approval for what was, unbeknown to a world dazzled by Soviet space spectaculars, a flagging programme unlikely to honour all the promises then being made by top design teams all vying with each other for approbation. By late 1963 the military, exasperated by soaring funds for space projects, were clamping down on the excesses they saw as diminishing the military balance.

Korolev, burdened by repeated Vostok flights, and wrestling with significant underfunding on the massive N1 rocket – Russia's long-term prospect for massive space projects – was keen to get a start on the mid-term prospect, Soyuz. But pressure of expanding work commitments meant that development of the 9K rocket stage had to be handed to Viktor P. Makeyev's SKB-385 at Miass, while development of the 11K tankers was given to Mikhail F. Reshetnev at OKB-10 located in Krasnoyarsk-26. Decentralisation was the order of the day, with design teams competing to out-promise their competitors.

The period between early 1961 and late 1963 was crucial for the Russian space programme, top secret and planned at the highest levels in closed chambers with few privy to the highest echelons in the Kremlin. Korolev had the ear of Khrushchev, who had trusted him to out-perform the Americans, but Chelomei employed Sergei Khrushchev, the premier's son, and had the other ear of the charismatic Soviet leader. Resisting pressure from the military for more attention to tanks, guns and aircraft, Khrushchev and his close supporters had been seduced by the global recognition that fed back to the Soviet Union from world tours conducted by its jubilant cosmonauts.

It was just this unexpected value in space achievement that propelled President Kennedy to launch the Apollo programme in May 1961

LEFT A KH-8 Gambit photo-reconnaissance satellite took this image looking down on the N1 launcher on its pad at the Baikonur facility on 19 September 1968. *(CIA)*

BELOW A series of stills from a video film of the separation sequence for the upper part of the payload shroud covering the LOK Soyuz after it had left the atmosphere. *(David Baker)*

Video grabs showing the test separation of the upper L3 fairing from the LOK

ABOVE In the complex sequence of flights necessary to assemble the early mission profile of the N1, the Soyuz launcher would still have had an important role to play in lifting the tankers and crew ships to Earth orbit. All that was eliminated when the decision was made to simplify the mission design and place all the elements together on the N1. *(NASA)*

as a Moon-landing mission to be accomplished by the end of the decade. As a direct reaction to Yuri Gagarin's flight on 12 April, it set in motion a massive expansion of US space activities far wider than the limited objective of Apollo. It was so grand a venture, calling for massive rocket power and big spacecraft beyond anything approved at the time, that new and ambitious plans were laid for testing the techniques that would be crucial for getting to the surface of the Moon.

NASA's tiny Mercury spacecraft was incapable of orbit changes and could sustain a single occupant for little more than a day in space. Apollo would require multiple orbit changes, development of rendezvous procedures, evaluation of docking techniques, qualification of space systems capable of sustaining three men for 14 days, and practice with working outside in a spacesuit. To do all that and prepare for Apollo, which was expected to start carrying astronauts on test flights by 1967 at the latest, would be an interim vehicle called Gemini.

As a stretched version of Mercury, built by the same manufacturer, it could be ready before Apollo and get a head-start on conducting all those new techniques and capabilities required of a mission to the Moon. As an interim vehicle for giving astronauts experience with space flight, Gemini was a well-publicised programme formally announced in December 1961. While the Americans knew almost nothing about the Russian space programme, not even the shape or design details of Vostok and its launcher, the Soviet intelligence machine was gathering a lot of information about Gemini and plans for Apollo.

In an environment in which Soviet design bureaus were continually falling in and out of favour, organisations coming and going unpredictably, the degree of competitive infighting would have been a shock to Western notions of a centralised bureaucracy run by an autocratic state. Conversely, it was in America that the centralisation they assumed for the USSR was rigidly and methodically making great gains on the competition. While Korolev was secretly trumpeting his massive N1, NASA had formally approved development of the Saturn V rocket that would carry astronauts to the Moon. This was a rocket which at inception

LEFT The development of the Soyuz spacecraft quickly focused on its role for Earth orbit rendezvous and docking and on its function for circumlunar and Moon landing missions, the basic three elements having been decided upon for all except the circumlunar flights. *(David Baker)*

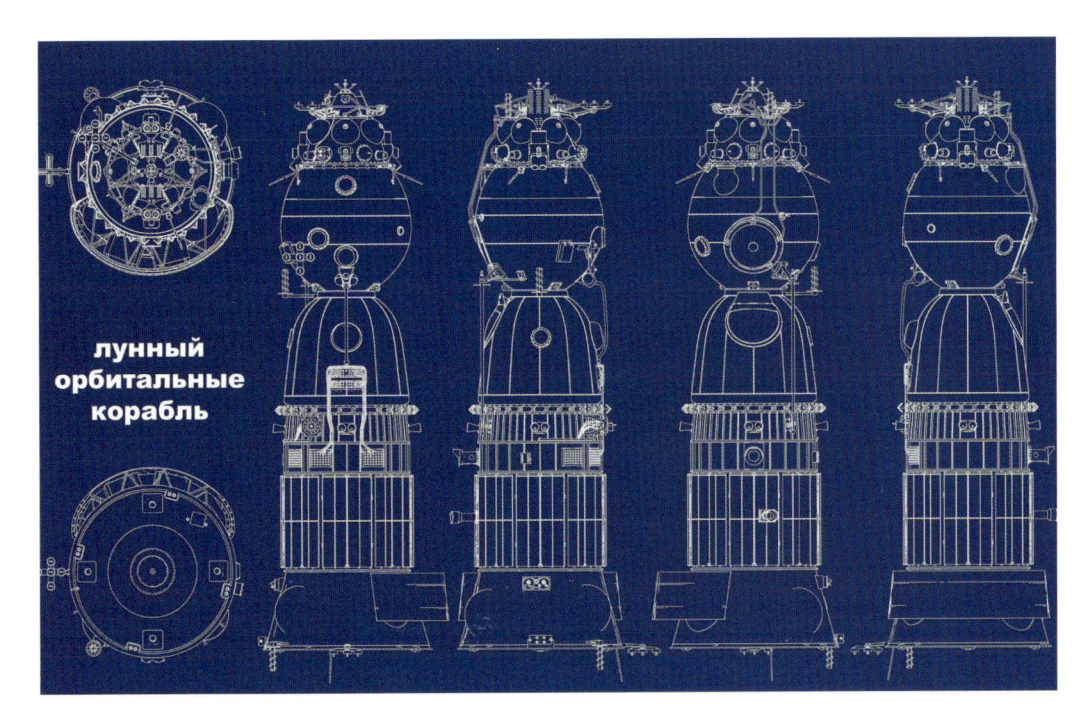

LEFT Blueprints of the LOK Moon landing mother ship show front, rear and side views at 90° rotation, with the complex rendezvous and docking equipment on the front face and the pronounced skirt at the base. *(David Baker)*

лунный орбитальные корабль

was capable of carrying into orbit a load greater than that projected for the later, developed version of Korolev's N1.

In July 1962, NASA decided on the method it would use to reach the Moon, opting for lunar orbit rendezvous, where a second spacecraft would descend from lunar orbit to the surface with two astronauts, returning to the mother ship for the flight back to Earth. This required a single rocket and two spacecraft to reach the surface, rather than three separate vehicles and six launches with Korolev's circumlunar mission.

By early 1963 NASA had selected the contractor for the second spacecraft, the lunar module, and the final piece of planning was in place. But nobody in Russia appeared to believe it could happen as advertised by the Americans, and planning for a flight around the Moon seemed to preoccupy Soviet thinking. While none of the internecine strife stalling Soviet plans was known to the Americans at the time, the only race in which the Russians focused their attention was the circumlunar flight. There simply were no design proposals or suggestions for a spacecraft capable of landing cosmonauts on the surface of the Moon.

So it was that Russia's rocket teams were focused more on the immediate future balance between their own capabilities and America's emerging Earth orbit programme in the form of the forthcoming Gemini missions. Because

BELOW The LK lander was a simplified design for one cosmonaut to go down to the surface of the Moon. Key: 1 Kontakt docking system, 2 attitude control thrusters, 3 rendezvous window, 4 viewing window for landing, 5 radar system for descent rate and altitude data, 6 descent thrusters, 7 landing gear, 8 communications antenna, 9 rendezvous radar for returning to the LOK spacecraft, A pressurised cabin, B pressurised equipment compartment, C hatch for descending to the lunar surface, D electrical power system and batteries, E ascent propulsion unit, G ladder for cosmonaut to step on to the Moon. *(David Baker)*

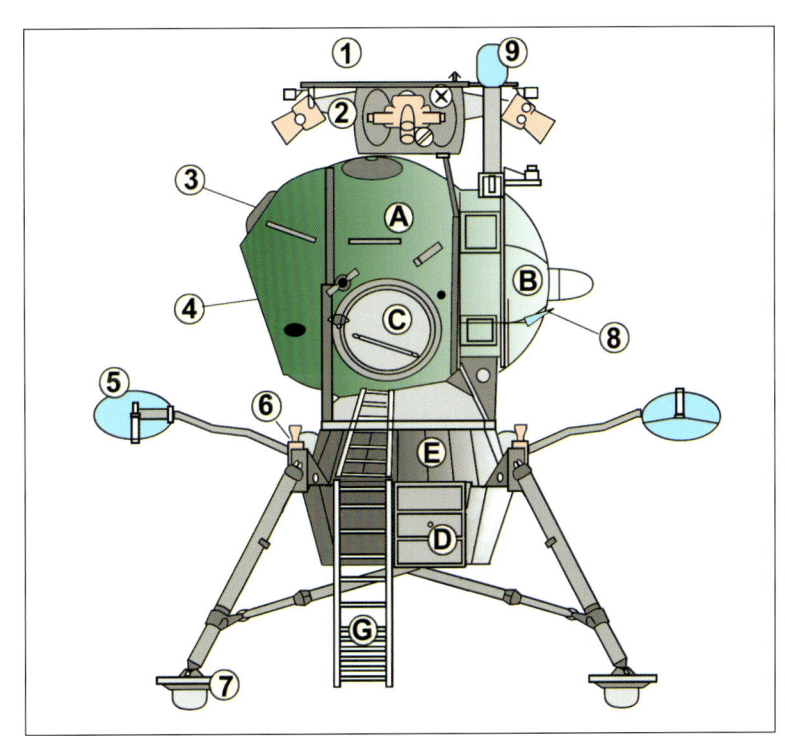

59

THE 7K-OK SOYUZ

LK Russian Lunar Landing Vehicle

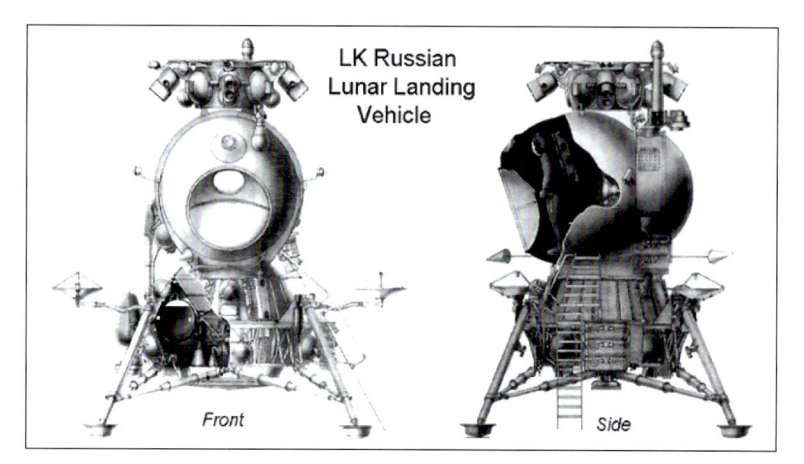

Front *Side*

ABOVE The general layout of the LK lander showing the small amount of space afforded the lone cosmonaut. The lander would have been decelerated to a very low descent velocity by a separate rocket stage jettisoned before touchdown. *(David Baker)*

of concerns over the rapid rise of Gemini, Chelomei staked his reputation on demolishing Korolev's circumlunar plan with its combination of 7K/9K/11K spacecraft and rocket stages and its multiple rendezvous operations. Instead, he proclaimed that his powerful UR-500K, still a very long way from test flights, could do the job in one launch, carrying a three-man crew around the Moon and back.

Chelomei proposed a spacecraft designated LK-1, configured much like Apollo but with features similar to Gemini, one or two crew members facing forward inside a pressurised conical compartment doubling as a re-entry vehicle. At the back, a service module would carry all the support systems and two solar array wings, much bigger in area than the arrays on Korolev's 7K/Soyuz and providing higher levels of electrical energy.

The habitable module would have a diameter of 9.2ft (2.8m) and weigh 8,380lb (3,800kg) attached to the cylindrical service module, both sections attached to the upper stage of the UR-500K. Weighing a total of 37,500lb (17,000kg), the LK-1 was to have been boosted out of Earth orbit by the terminal stage of the launcher and, after separating, use a main propulsion system in the service module for trajectory correction manoeuvres.

A 'go' for the Moon

By mid-1963 several competing factions were in play to delay any formal decision about the future of Russia's human space flight programme, not least among the rocket engineers themselves. At the heart of this conflict were Korolev and Valentin Glushko, a rocket engine designer who had developed many of the propulsion systems used in early Soviet rockets. But the two men were in deep disagreement over which type of rocket motor should be developed, a confrontation that went back to the early 1950s.

Each was highly opinionated and self-assured to the point of extending beyond their specialities, each intruding into the professional domain of the other, causing acrimonious insults to fly. Korolev was in deep argument over the propulsion systems for his rocket/spacecraft combination, and because Khrushchev was caught up in the inertia of his beloved space programme he saw at close hand the

RIGHT The relative sizes of Russia's one-man LK and NASA's two-man Lunar Module show not only the difference in size but also point to the limited capacity of the Russian vehicle to carry the additional equipment which would become standard on NASA flights to the surface. *(David Baker)*

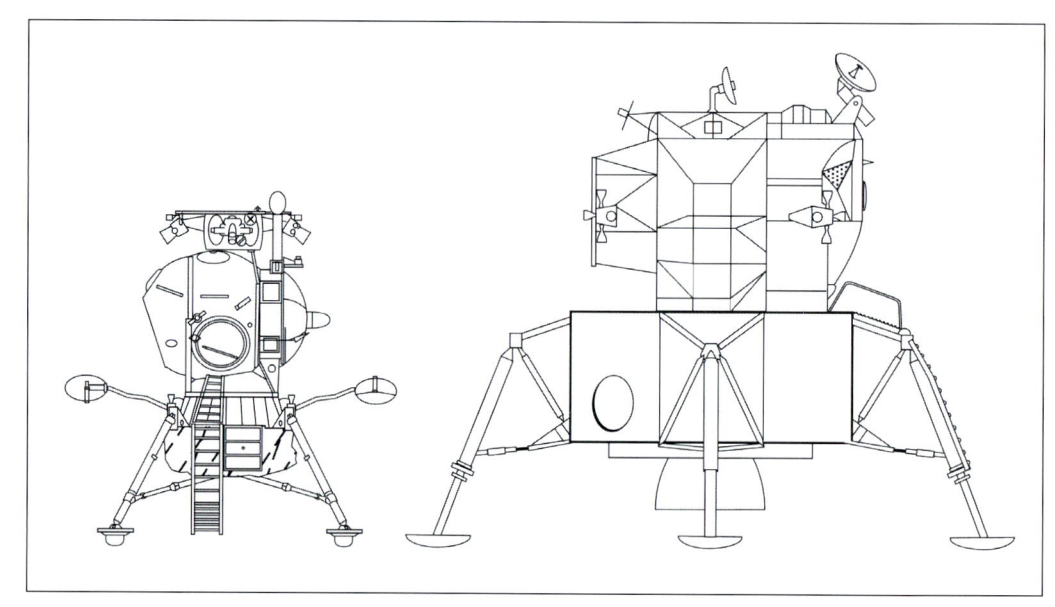

debilitating effect that this was having on decision-making.

Khrushchev invited Korolev and Glushko to his private dacha in June 1963 and tried to get the two men to settle their differences. Contrary to the situation in America, where the US government had set up a civilian space agency with focused responsibilities, the Russian space programme was an opportunistic flagpole for waving the hammer and sickle before the world community of uncommitted nations and had no centrally approved plan or mandate.

Everyone agreed that the space programme was useful propaganda, but irrelevant to the main body of work to which the Kremlin was committed: economic growth and a dominant military force. To some extent, Khrushchev was fighting for his political life, but the more he dug deeper to get increasingly extravagant and spectacular achievements into space, the more it sucked money from other government activities and rallied his opponents, tired of what they saw as digressions.

At the meeting between Korolev and Glushko, Khrushchev's appeal for the two men to settle their differences and combine energies on projects that could benefit the Soviet Union fell on deaf ears. In later testimony, Sergei Khrushchev said that he thought it was probably at this meeting that his father realised the impossible challenge that had been laid down by the Americans. For, while the rocket engineers were dismissive of the Apollo plan, promoting a circumlunar trip as an achievable goal, deep down Khrushchev knew that there was a very high probability that the Americans would achieve their goal – and on time.

When he asked for detailed plans about a possible Soviet landing on the Moon, Khrushchev was told it would take three N1 launches to assemble a giant spaceship to get to the surface. And when he asked about the cost and was told it would be in the range of 10–12 billion roubles, the reply seemed to tip Khrushchev away from the idea. In currency values of the time, this was equivalent to $11–13 billion, or $83–98 billion in 2014 rates. But interest had already been sown, and elsewhere there was considerable pressure to approve not only a circumlunar trip but a landing too.

Responding to what he perceived as an enthusiasm for ambitious space projects, and so as to assume a leadership role in this hotly contested area of Soviet technological virility, Korolev laid down a schedule of five major programmes. These included his circumlunar flight (L1); a 5.5-ton lunar roving vehicle using the 9K/11K stages (L2); a Moon landing with the N1 rocket launching a 7K/Soyuz and a manned lander (L3); a lunar orbit flight for scientific analysis of the Moon using the N1 and a modified Soyuz (L4); and an advanced roving vehicle launched by the N1 capable of carrying cosmonauts far across the surface (L5).

Contrary to this single-minded approach from the space engineers, Khrushchev responded to a most unexpected message from the United States. It was an offer from President Kennedy for the US and the USSR to foreclose the threat of a runaway space race and merge their goals. Believing, inaccurately as it turned out, that Russia was driving headlong for the Moon, Kennedy was reacting to a gathering discontent with the escalating cost of America's own highly ambitious space programme.

In May 1961 Congress and most of the American public had been prepared to accept Kennedy's word that a bold and challenging new initiative was required to restore US pride in itself, not least to regain the confidence of a world which now seemed more inclined to believe that Russia was the next great superpower. All the intelligence reports back from the US Information Service said so. Two years on from that challenge being thrown down, it did not seem so wise an idea. The true cost of getting to the Moon was beginning to escalate, and there were murmurings of discontent among the political elite who had approved the money for Apollo.

Kennedy was so concerned over the cost that he quizzed NASA boss James E. Webb on the magnitude of the programme, saying that he was not interested in a big space programme, only in beating the Russians to the Moon. While the Soviet leadership were not at all convinced that such an ambitious goal was achievable within the time set, Russian rocket engineers knew only too well that what the Americans set their minds to would most likely be done. Only, it seemed, was the political leadership in both countries out of alignment

LEFT The general design of the LK lander was capable of remaining on the surface for 24 hours and had little flexibility for additional time or for carrying extra equipment. *(David Baker)*

with the expert opinion of their space scientists and engineers.

When public opinion polls showed that the majority of Americans disapproved of the Apollo objective, and realising that by starting a series of flights with the giant Saturn I in October 1961 America had already drawn level with Russia, Kennedy looked for a way out. In fact, the increase in space spending since 1960 had been greater than for any other government activity in peacetime, rising from $401 million to $744 million in 1961 to $1,257 million in 1962, $2,552 million in 1963, and $4,170 million in 1964, this last period covering the year beginning 1 July 1963. The following year, NASA would receive $5.1 billion.

Several back-door messages were sent to the Soviet premier and open discussion alerted the media to these thoughts, culminating in a speech to the General Assembly of the United Nations on 20 September 1963 in which Kennedy offered the Russians the possibility

RIGHT Examples of test and training hardware remain from the failed N1/L3 programme. This LK in a Moscow museum bears sad testimony to the failed ambitions of a brilliant team of engineers let down by the politics of the system. *(David Baker)*

FAR RIGHT The descent viewing window and the hatch through which a cosmonaut would descend to the surface are clearly visible on this LK mock-up. *(David Baker)*

of a joint flight to the Moon. Kennedy had already alerted Webb to his intentions and had asked the NASA boss to 'watch my back' on any repercussions from this startling move to radically change the agenda.

Whether it was this Cold War olive branch or a realisation that the Americans were indeed very serious, and fully committed to the Moon goal, or whether it was a determination at last to decide upon a landing mission in addition to the circumlunar flight, the mood in Russia changed dramatically in early 1964. Entirely coincidentally, on 23 September 1963 Korolev submitted his five programmes for expanded operations focused on the Moon.

But there was competition for Korolev from chief design engineer Mikhail K. Yangel of OKB-586, who was developing a heavy-lift rocket designated R-56. While smaller than Korolev's mighty N1, the R-56 was plausible as a contender for manned Moon landings, albeit with multiple launches, and it was more than a match for Chelomei's UR-500K. Yet none of these proposed rockets had approved missions, and Khrushchev's declining enthusiasm for the Moon race was sapping support – and funds. By early 1964 the N1 was going nowhere and still lacked a mission.

It was only when Korolev again met with Khrushchev, on 17 March 1964, that he extracted a commitment from the Soviet premier for a Moon plan. Korolev realised that the very survival of the N1 was contingent upon a role for it that was incapable of being satisfied by any other rocket. The two were inextricably linked. Even though a verbal commitment was given, again progress was slow, accelerated by sharp and alarmist messages coming from the Korolev team. In a letter dated 25 May 1964 a direct appeal was made to Leonid Brezhnev, Secretary of the Central Committee for Defence Industries, on the basis that it was now a matter of national urgency.

While the Americans were totally unaware of all this, the Russians were watching as NASA neared the first flights with astronauts aboard their Gemini spacecraft and made steady progress with flights of the Saturn I rocket carrying mock-ups of the Apollo spacecraft. Already, NASA was flying a highly efficient liquid hydrogen/liquid oxygen upper rocket stage on Saturn I, precursor to similar stages on the giant Saturn V. The Russians were many years away from achieving parity. In truth, Kennedy had been right when intimating that the race had already been won.

LEFT A portion of the lower section of the LOK spacecraft displays thruster locations and external equipment. *(David Baker)*

Long overdue, events now moved rapidly. On 19 June 1964 agreement was reached to increase funding for the N1 rocket, 16 being ordered for delivery between 1966 and 1968. On 24 July the Military-Industrial Commission sanctioned the five-point plan posited by Korolev, with the L3 mission selected for Moon landings to take place from 1968. In addition, in 1967 Chelomei's circumlunar mission would precede the landings to maintain what the Commission regarded as Russia's pre-eminent position.

In its original form the L3 concept was gargantuan in the extreme. A series of three Earth-orbit rendezvous flights with the N1 super-booster, carrying first the Moon lander and then two carrying propellant for the injection stage, would be followed by an R-7 derivative carrying a 7K/Soyuz to dock with the other elements. This total mass of 200 tons would include the lunar lander weighing 46,300lb (21,000kg). But this changed with the August decree, and Korolev rewrote the mission concept to involve only one N1 launch vehicle.

The timing for these events to occur was grossly over-optimistic and had no sensible chance of success, especially as interim Voskhod adaptations of Vostok had still to work their way through to flight status. The magnitude of the work involved seemed not to have been factored into the planning, but with the upcoming 50th anniversary of the Russian Revolution in 1967 the Soviet leadership were fearful of the Americans pulling off a circumlunar flight first and even more worried than they would miss this historic date for their own space spectacular.

On 3 August 1964, Korolev's 7K/9K/11K circumlunar plan was formally abandoned in the wake of the decision to give that work to Chelomei and his one-man LK-1 spacecraft. Under the revised structure, Korolev would retain his work on Voskhod, still hope to develop the Soyuz spacecraft and produce the N1 rocket, as big as NASA's Saturn V but with less payload capability, for the Moon landings to follow. Work on Yangel's R-56 had been cancelled on 19 June.

Redirection

Korolev's L3 Moon-landing proposal was now to involve only one N1, the multi-launch scheme having been replaced with a lunar-orbit rendezvous method identical in principle to that adopted by NASA in 1962 for its Apollo programme. Now known as the N1-L3, it would task the capabilities of the booster rocket and bring a plague of problems that never would be fully resolved. With a lifting capacity of 75 tons to low Earth orbit, Korolev was convinced he could design a pair of spacecraft to carry out the mission within that limit, by trimming back on excess weight and improving the performance of the N1.

With Chelomei busy developing his LK-1 one-man circumlunar Moon vehicle, Korolev set about adapting his 7K/Soyuz for a different role to that for which it had been proposed – his own 7K/9K/11K circumlunar mission. A guiding light who played a major role in reshaping Soyuz, and a prominent figure in Korolev's design team, Boris Chertok, would play a strong role in reconfiguring Soyuz. Another key member of the Korolev team was Konstantin D. Busheyev, who had nursed the 7K/9K/11K project from its inception. Now these men and

RIGHT The rear face of the LK reveals a complexity of pipes, conduits and tubing that would have been covered with thermal insulation on a flight-rated spacecraft. *(David Baker)*

many more were tasked with finding a role for Soyuz to save it from extinction.

Much work had been done at several institutes and research centres honing the definitive design configuration of this vehicle. Korolev was not about to have it evaporate before his eyes. But he lost a valiant protector when Nikita Khrushchev was ousted from power in a bloodless coup on 14 October 1964. Nevertheless, the work had already been done and approval for this revamped march into the future had already been received.

Much as Voskhod had been adapted from the Vostok, so would the Soyuz eventually be built as an outgrowth of the Sever, L1 and 7K concepts for which so much work had been done over the preceding four years. And it was in the legacy of this earlier work that the Korolev team found a functional purpose for the 7K/Soyuz, for which they received approval in February 1965. Korolev successfully argued that as Soyuz had been designed for rendezvous and docking missions, and that as the lunar-orbit rendezvous mode had finally been selected for his N1-L3 Moon landing mission, the use of Soyuz to practise and rehearse these techniques around Earth would be a useful precursor to the Moon mission itself.

In this way he was arguing for a role NASA had charged to the Gemini programme, paving the way for the bigger missions to come. The logic was self-evident. Two Soyuz spacecraft could evaluate rendezvous equipment vital to the N1-L3 mission, conduct docking operations with each other and demonstrate the various orbital techniques necessary for re-rendezvous and re-docking. In effect, conducting a complete profile of lunar mission manoeuvres, but in the relative safety of low Earth orbit, close to a dash for home should anything go wrong with either spacecraft.

In early 1965 Soyuz development shifted to Department No 93 at OKB-1, and to define its new function as a test vehicle in orbit it was designated 7K-OK, the 'OK' being Russian for 'orbital ship'. In tidying up the associated work, the work orders for the 9K and 11K tanker-stages were cancelled and the new Soyuz given a line number: 11F615. Very little change was required in the general configuration of the spacecraft. It still retained the three essential elements of an orbital module at the front, a descent module in the middle and an instrument module at the back. But a lot of the associated systems and subsystems were tailored to the new role it was given.

LEFT Displaying the meaning of a 'minimum-sized vehicle', the LK landing gear would have been folded against the lower section until separating in lunar orbit for the descent phase. *(David Baker)*

The general specification given to the 7K-OK was quite basic: it was required to remain in space for up to ten days and total mass was not to exceed 14,470lb (6,560kg), and it was to adopt rendezvous and docking equipment relocated from the rear to the forward end of the vehicle on top of the orbital module. The spacecraft itself would be encapsulated on top of the second stage of the rocket by a shroud that would be jettisoned before reaching orbit. Most challenging of all was the tandem launch escape system, introduced for the first time to a Russian manned spacecraft, where the spacecraft would be lifted free of the ascending rocket should anything go wrong to threaten the lives of the crew. Most challenging of all were the many complex systems required to carry out its advanced mission profile.

The 7K-OK programme would be notable for several firsts, another of which was the development of a launch vehicle uniquely

ABOVE In much the way that NASA's Lunar Module had a descent stage from which the crew would depart for Moon orbit, so did the Russian LK have its ascent section lift free from the landing base. Note that with the LK the propulsion system used for the final deceleration before touchdown would also be used for ascent. *(David Baker)*

designed for this particular spacecraft. Known officially as the 11A511, it was a derivative of the 11A57, and a development of the same line of launch vehicles that had been used for the Vostok and Voskhod programmes. But they had adopted upper stages to R-7 which had been developed for other, military, programmes involving satellite launches, and for science missions to the Moon and the planets. And, just to add confusion, because it was unique to the spacecraft the launch vehicle too would be known as Soyuz – right up to the present.

The 11A511 adopted the same upper stage as that used for the Voskhod flights, but with the more powerful RD-0110 delivering a thrust of 67,000lb (298kN) instead of the RD-0108, with a thrust of 66,100lb (294kN) on the 11A57. Several major changes had been made to this vehicle, elevating it to a more efficient mode of operation than its predecessors. The telemetry equipment had been reduced in weight to less than 331lb (150kg) and the rocket motors were selected by hand for their high standard of manufacture and performance when tested prior to being installed in the stages. The four boosters each had an RD-107-11D511 while the core stage had the RD-107-11D512. The launch vehicle and spacecraft had a pad weight of 679,000lb (308,000kg) at launch and an initial lift-off thrust of 907,700lb (4,037.7kN). It had a height of 149.6ft (45.6m) to the top of the launch escape system.

The plan was to have two 7K-OK spacecraft ready by the end of 1965, two more in the first quarter of 1966 followed by three more in the second quarter. Various tests including air-drops and parachute evaluation would be completed, along with qualification that the spacecraft could survive a water landing and remain afloat. The ambitious flight plan called for the first docking of two Soyuz spacecraft in early 1966, using an automated system that was causing concern. Some engineers wanted the cosmonauts to have a sophisticated manual backup rendezvous and docking system, but this was brushed aside by Korolev, who was convinced that the new automated system would be sufficient.

The redirection of the Soyuz programme would see it gradually move to centre-stage among the several different spacecraft and launch vehicles being proposed. The rendezvous and docking role for Soyuz was crucial to Korolev's bid to seize control of the circumlunar mission from Chelomei. Korolev achieved a major coup over his arch-rival on 25 October 1965 when it was decided that a version of the 7K-OK known as the 7K-L1 would replace Chelomei's L1 spacecraft. No longer would it be a Gemini-style precursor to Moon missions, solely responsible for testing rendezvous and docking techniques in Earth orbit.

The Soyuz variant would still be launched to the Moon on Chelomei's UR-500K, but initially to get there it would first be placed in Earth orbit, where a standard 7K-OK would dock with it and transfer the crew, which would then depart for a flight around the Moon. In the process of refining this mission plan, the two flights were compressed into one flight, the 7K-L1 being launched with its crew on board, obviating the need for a separate launch.

The circumlunar 7K-L1 was a stripped-down 7K-OK with reduced weight to keep it

ABOVE Departing for home, the LOK mother ship brings its crew back to Earth at the conclusion of a Moon landing. *(David Baker)*

within capacity of the UR-500K, a maximum 12,100lb (5,500kg) being allowed – some 440lb (200kg) lighter than the Earth orbit version. But the marginal difference in weight belied the significant change in configuration, since the 7K-L1 would not carry the spherical orbit module for the trip around the Moon and back.

Plans for flights of Soyuz spacecraft by the end of 1965 were overly optimistic and the inevitable delays piled up, shifting the debut of this multi-seat spacecraft to the following year. After Korolev's death in January 1966, his successor Mishin pressed ahead with development of engineering systems for Soyuz, focusing resources by stopping all further plans for Voskhod missions and scheduling a series of test flights toward the end of the year.

The initial direction of Soyuz had tilted to directly support the circumlunar and Moon landing missions and as precursor to the 7K-L1, which would be numbered within the Zond system when flights around the Moon began. By mid-1966 a preliminary plan had emerged for two automated orbital tests with the 7K-L1 followed by two automated circumlunar flights, all four by the end of the year. These would be followed by five manned circumlunar flights between December 1966 and May 1967 using separately launched 7K-OK spacecraft to bring the crew to Earth orbit prior to departure. Following those, by September 1967 there were to be four manned circumlunar flights with the crew aboard the 7K-L1 spacecraft at launch.

This highly optimistic schedule was doomed to change and by late 1966 Mishin had drastically reduced the flight schedule, preceding a manned circumlunar flight planned for mid-1967 with four automated test flights, beginning with a mission to Earth orbit for a test of rendezvous and docking techniques. When the engineers complained that even this was a wildly over-optimistic proposition, Mishin merely repeated what he had been directed to achieve by his political masters – that the USSR must celebrate the 50th anniversary of the October 1917 revolution with a manned circumlunar mission.

Feature A

The Soyuz spacecraft

From the outside, the Soyuz spacecraft appears to have three separate modules attached together in a tandem configuration. In fact there are five, consisting of three pressurised and two unpressurised segments.

Manufactured from magnesium alloys (later versions in aluminium), the pressurised orbital module (OM), sometimes referred to in training documents as the 'habitable module' or the 'crew rest module', is at the front. It is 8.5ft (2.6m) in length with a diameter of 7.2ft (2.2m) and weighs about 2,865lb (1,300kg), with a pressurised volume of 229.5ft^3 (6.5m^3) and a habitable volume of only 141ft^3 (4m^3). It contains equipment and apparatus necessary for a range of spacecraft systems, including life support for the three-person crew, rendezvous and docking equipment, control panels for systems operation, measuring devices for the mission, TV systems and crew rest facilities.

The OM will remain attached to the Soyuz throughout its orbital operations and will only be separated prior to the crew returning to Earth. It incorporates docking equipment at the top within which is set a hatch, 31.5in (80cm) in diameter, for crew transfer to the International Space Station, set within the docking ring which connects it to the ISS docking port. At the opposite end of the OM is another hatch, 23.6in (60cm) in diameter through which the crew pass to enter the descent module (DM) below for launch.

The upper part of the OM contains a window that can be used by the crew for manual docking operations should the automated system fail. On the lower side are special umbilical connections carrying conduits and electrical wiring between the two modules. On the exterior of the OM, antennae for the Kurs radar system facilitate automatic rendezvous and docking, which has become a speciality of Soyuz spacecraft. Pyrotechnic locks and spring-release systems hold the two modules together during flight. The OM is attached to the DM, the pressurised compartment where the crew reside during launch, docking, undocking and re-entry operations.

The DM is in the shape of a truncated cone, having the appearance of a headlight fixture on an old car. It has a pressurised volume of 136ft^3 (3.85m^3) and a habitable volume of 88ft^3 (2.5m^3). It has a height of 6.9ft (2.1m), a diameter of 7.2ft (2.2m) and weighs 6,400lb (2,900kg). The cramped quarters provide contoured couches for each cosmonaut in the crouched position in front of a crew display and controls panel.

The DM is designed for land recovery but

BELOW The Soyuz spacecraft of today is the result of two major evolutionary steps from the original design first flown unmanned in November 1966. The first step was in 1971 when the spacecraft was adapted into a ferry vehicle for space stations, adding pressure suits for the crew and focusing technical developments on the reliability of spacecraft systems. The second step occurred with the introduction of the T-series in 1978 with digital computers, an integrated fuel supply system and a redesign of the interior cabin layout. All successive versions flow from that. *(Energia)*

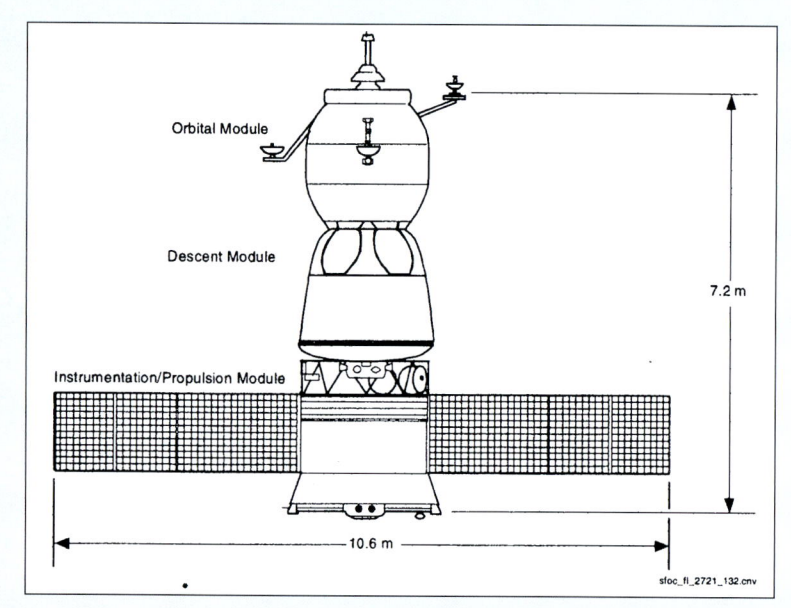

is capable of remaining afloat should it land in water. It is also a shelter where the crew can remain in harsh weather conditions, such as can be experienced during winter when snow and blizzard conditions have been known to delay the arrival of recovery teams. The DM contains controls and equipment for managing space flight operations, for essential life support after jettisoning the attached modules fore and aft, and post-landing survival kits.

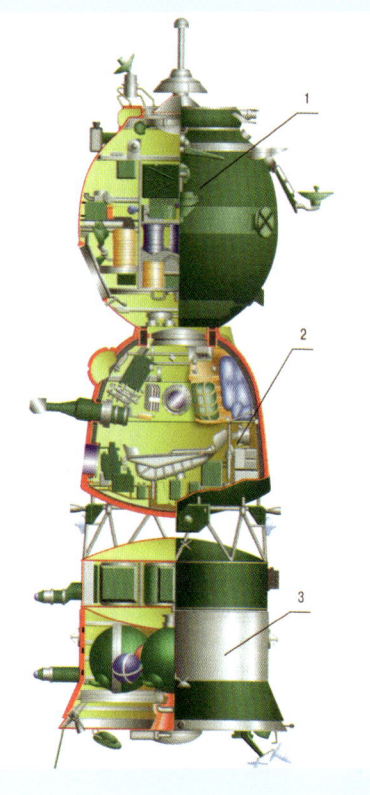

It also has TV and communication systems, electrical power supplies and a telemetry and data transmission system. The three shock-absorbing seats are designed to rise on struts shortly before landing to cushion the impact of striking firm ground. The exterior of the DM is covered with thermal protection and the base of the semi-conical structure, which forms a modest convex profile, is jettisoned during descent through the atmosphere to expose

LEFT The general configuration of the Soyuz consists of three elements: the orbital module (1) on top, the descent module (2) beneath and the instrument module (3) below. Passive thermal protection covers the exterior of the pressurised sections, with the propulsion section of the instrument module in white. The launch vehicle adapter is at the bottom. *(Energia)*

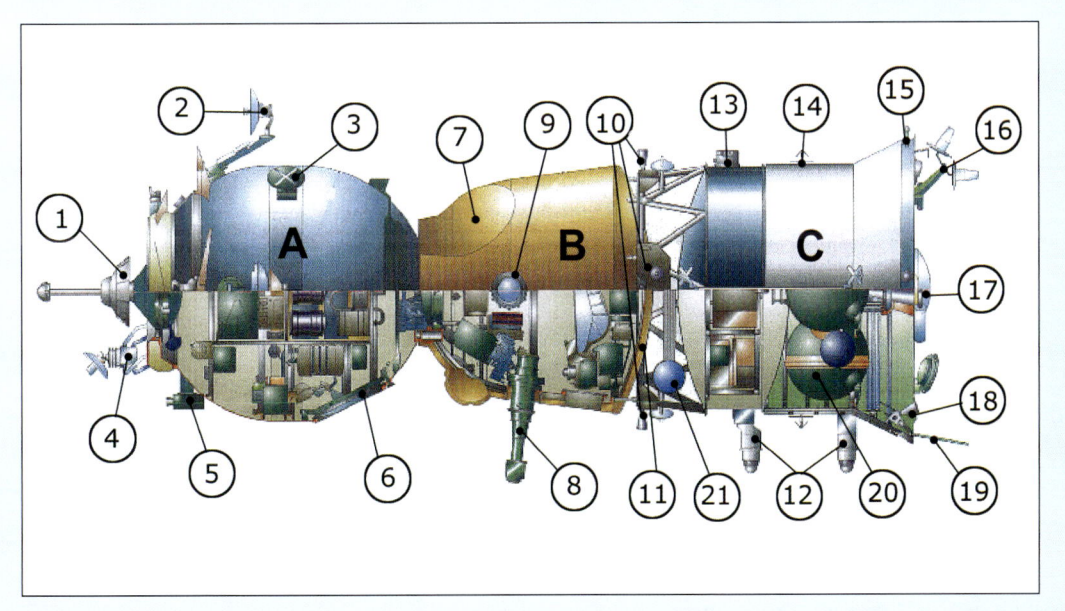

LEFT Key: orbital module (A): 1 docking mechanism; 2 & 4 Kurs antenna; 3 TV antenna; 5 camera; 6 hatch; descent module (B): 7 parachutes; 8 periscope; 9 window; 11 heat shield; instrument module (C): 10 & 18 thrusters; 12 Earth sensors; 13 Sun sensor; 14 solar panels attachment; 15 thermal sensor; 16 Kurs antenna; 17 main propulsion system; 19 communications antenna; 20 propellant tanks. *(Energia)*

THE SOYUZ SPACECRAFT

ABOVE **The design of the Soyuz has rendered it suitable for adaptation into an unmanned Progress cargo tanker. In the foreground, Soyuz TMA-9 partly obscures Progress 22 docked to the International Space Station. This flexibility extends to the use of the instrument assembly as the means of transporting airlock modules to the ISS.** *(NASA)*

thrusters designed to fire milliseconds before impact to further decelerate the spacecraft.

Special pressurised containers in the DM contain primary and backup parachutes, with two windows provided for exterior views. Eight reaction control thrusters are situated on the outside of the DM for attitude control during descent outside the atmosphere, their toxic hydrogen peroxide propellants being vented as the spacecraft enters the denser layers of the atmosphere. The top hatch has a diameter of 27.5in (70cm) and is used by the crew to reach their couches during ingress prior to launch, to move in and out of the orbital module during flight, or to egress their spacecraft after landing.

An open truss structure supports the descent module and is attached to the instrument/assembly module below. It contains separation equipment with some attitude control thrusters, radio antennae and electrical connectors for ground equipment before launch. It also contains oxygen tanks for the atmosphere revitalisation system. The truss structure is the interface between the pressurised crew and orbital modules above and the instrument, or service, module below.

The combined instrument/assembly module (IM) has a height of 8.2ft (2.5m) with a diameter at the top of 7.2ft (2.2m), where it is attached to the descent module truss, and a diameter of 8.9ft (2.7m) at the bottom, where it mates to the upper stage of the launch vehicle. The upper part of the drum-shaped IM is pressurised with an inert gas and has racks of equipment including radio systems, telemetry equipment, power supply modules, thermal control equipment and attitude and orientation modules.

It also contains infra-red and solar attitude sensors, attachment points for the pivoting solar cell arrays and conduits for transferring power, fluids and data across the truss structure and into the descent module for crew and life support. Folded against the side of the spacecraft for launch, when deployed the solar arrays have a span of just over 35ft (10.7m). Various configurations of solar arrays have been used in different versions of Soyuz over time. Initial versions of Soyuz had four panels with a total surface area of 150.5ft^2 (14m^2) but different arrays were used with three sections. The TM and TMA configurations use four-segment arrays.

Early Soyuz vehicles carried solar array wings but these were deleted in versions used after the introduction of the spacecraft to the ferry role, carrying crew members to and from the Salyut space stations. In this 7K-T variant, the spacecraft relied exclusively on battery power, a mission length of three days being sufficient for a two-day rendezvous profile and a single day to return to Earth. This was the version introduced after the loss of three cosmonauts aboard Soyuz 11 in 1971, adapted to carry two crew members wearing Sokol pressure suits.

In Soyuz T, TM and TMA variants, the solar cells were restored. Reintroduced on Soyuz T, the solar arrays were smaller and more efficient than those used on the first-generation Soyuz previously used for independent flight and for flights to the Salyut 1 station. With Soyuz TM, the spacecraft got triple redundant electrical systems and a considerable upgrade in the way the power was distributed throughout the vehicle. There were also a greater number of redundant electrical switching options, which greatly enhanced the operability, and potentially the reliability, of the later series, including the TMA.

Under optimum conditions the solar arrays on the TMA generation provide an optimal power level of 85A at 23–34vdc. After 180 days in space, the battery capacity is 100Ah. On the

RIGHT Departing the ISS, TMA-19 displays its solar array wings, which have been redesigned several times and are now characteristically flat with each wing consisting of four panels. Note the uniquely Russian docking mechanism on the front of the orbital module. *(NASA)*

first revolution after docking with the ISS, the spacecraft is configured to combined power supply mode, where the electrical system is hooked up to the station, only returning to autonomous power supply mode when preparations are made to undock and de-orbit. The design of the electrical power system has biased autonomy, with several automatic switching options controlled by the central electrical systems processor. Voltage and current indicators provide the crew with visual indications of the state of the electrical systems.

The lower segment of the cylindrical IM comprises the unpressurised propulsion section with propellant tanks, attitude control thrusters, the main propulsion system engine and radiators with a surface area of 86.1ft² (8m²) for dumping excess heat into the vacuum of space. On the exterior, the module has communications antennae and additional attitude control system sensors. The IM has a weight of approximately 5,700lb (2,600kg).

Over the years several modifications have changed the specifics of the spacecraft, and with modifications and upgrades Soyuz is in a continuously evolving form. Most of the changes have been internal, with considerable changes to the electronic equipment, sensors and other mission-enabling systems. Overall, the Soyuz TMA spacecraft has a height of 23.6ft (7.2m) and a total weight at launch of up to 15,875lb (7,200kg), precise values depending on the mission.

Atmosphere and water supply

Habitability aboard Soyuz is supported by the atmosphere revitalisation system (ARS) which provides life support for the crew of two or three cosmonauts in a 4.2-day flight, of which 2.2 days is for ascent to the International Space Station and two days is for descent. Flight controllers have since introduced a fast ascent with launch to docking within one day.

The environmental requirements are, therefore, considerably in excess of the periods usually used for manned Soyuz flights. In addition, provision is made for life support during an emergency when the descent module has lost pressure and the crew are surviving in their Sokol spacesuits, for which they have 125 minutes of oxygen – enough to go through an emergency descent and landing.

Equipment for maintaining a habitable environment in the DM is designed for pressure to be contained within a band of 8.7–18.7lb/in² (450–970mm Hg), with a nominal pressure maintained at an average 14.7lb/in² (760mm Hg). Pressure readings are divided into two regions of 8.12–13.1lb/in² (410–690mm Hg) and 13.9–19.1lb/in (720–990mm Hg). Between these two bands, there are 20 pressure-reading positions with a total sensor pressure error band of +/- 0.58lb/in² (30mm Hg).

An equalisation valve is fitted to maintain pressure at the desired level and a pressure relief valve is installed to prevent over-pressurisation. A squib valve explosively

BELOW The orbital module has undergone several subtle changes over time, the one permanent option being the type of docking mechanism carried forward of the front hatch. Access to the descent module is made via a hatch in the side of the OM. *(Energia)*

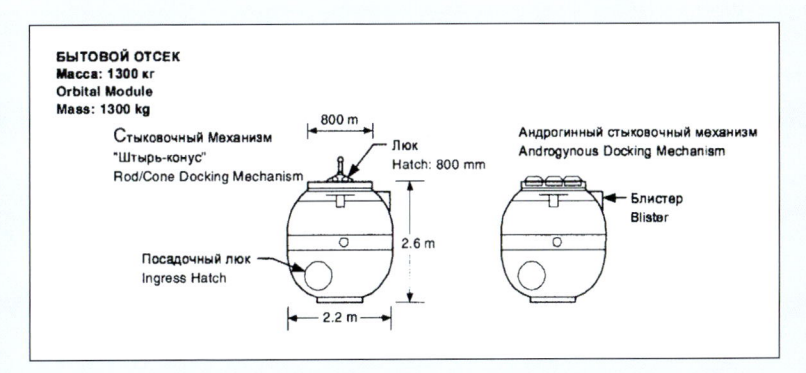

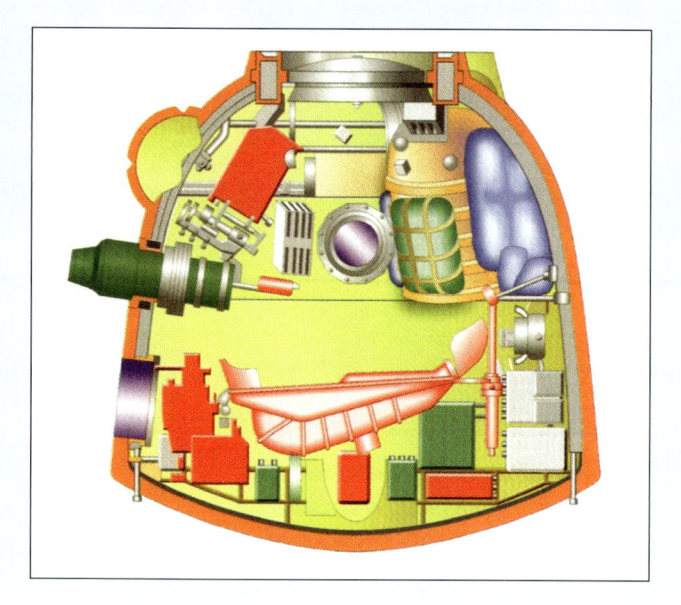

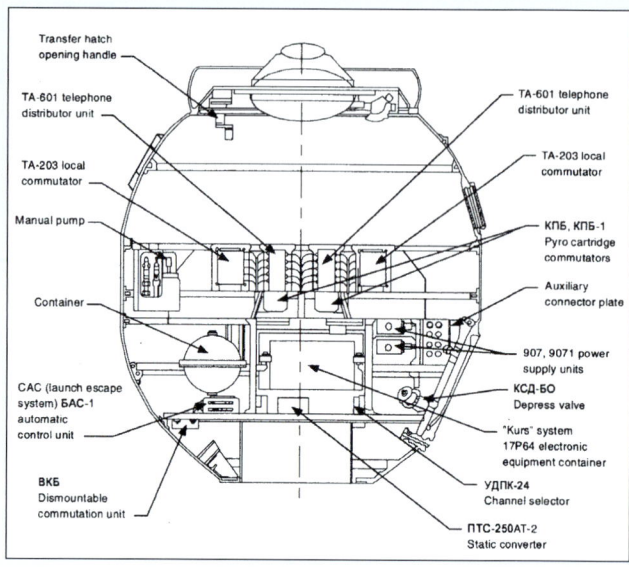

ABOVE The interior layout of the descent module has changed little in general configuration and is characterised by the position of the couches, the flat forward display and instrument panel with the optical device and periscope penetrating the exterior wall of the module. *(Energia)*

ABOVE RIGHT One half of the orbital module interior showing the equipment revealed when the front-facing panel is removed. *(Energia)*

depressurises the cabin in the event pressure increases beyond 18.4lb/in² (950mm Hg) and a locking valve closes when pressure falls below 10.6lb/in² (550mm Hg). There is also a pressurisation system installed in the parachute compartment of the descent module to ensure it remains afloat in the event of a water landing.

The air temperature is maintained in the range 64–77°F (18–25°C) with a humidity of less than 75%. The systems and equipment for the life-support function are rated for a life of 187 days, comfortably within the 180 on-orbit design life of the Soyuz spacecraft. This has been considerably extended over time, with Soyuz TMA spacecraft rated at 240 days, but earlier Soyuz vehicles had an autonomous life measured in days, although they were capable of being stored in dormant mode docked to Salyut space stations.

The environmental systems are based on the human requirement for 25 litres/hr of oxygen and the excretion of 20 litres/hr of carbon dioxide. The system is also configured to prevent the partial pressure of carbon dioxide exceeding 0.2lb/in³ (10mm Hg) and to keep water vapour below 0.3lb/in³ (15mm Hg). The CO_2 purification cartridge is based on a superoxide that releases lithium hydroxide to absorb carbon dioxide and plays the same function as similar systems on earlier spacecraft. Each cartridge has a 'life' of 60 man-hours and is replaced when the CO_2 level reaches the specified limit of 0.2lb/in³. Flight plan instructions demand a new cartridge is fitted for each of three critical phases: orbital injection, orbital flight, and docked flight. Crew members are required to change the cartridge on these cycles or on the service life defined by the specification.

The water supply system includes a storage tank, manual pump, drinking dispenser and individual mouthpieces for each crew member, this equipment being installed in the orbital module opposite the ingress hatch. The stowage tank has a capacity of 5.28 US gallons (20 litres) with a feed capacity of 0.22 US gallon (0.85 litre), this being useable after 360 days in the stowage tank and 120 days in the feed tank. The assumed rate of water consumption is 0.45 US gallon (1.7 litres) per person/per day.

Food provision includes daily rations that provide 3,000kcal/person, with specific packages coded for each crew member. Three meals a day are assigned to each cosmonaut, with associated tools including can openers and utensils. Human waste is contained in a solid/liquid waste collector, a ring collector for faecal material and a urine collector similar to a condom with inserts and replaceable rings. The life of the system is rated at 12.6 man/days, the collector having a capacity of 2.85 US gallons (10.8 litres) with each person excreting 0.32 US gallon (1.2 litres) per day.

Communications

The Soyuz TMA 17V12 Rassvet communications system is much improved over the T series but the basic characteristics and operating procedure remain much the same. It is based on a two-way duplex radio communication system in the VHF band, supplemented with a two-way simplex mode for communications between the spacecraft and the ground for contingencies and for communication with the International Space Station. The VHF system transmits at 121MHz and receives at 130MHz.

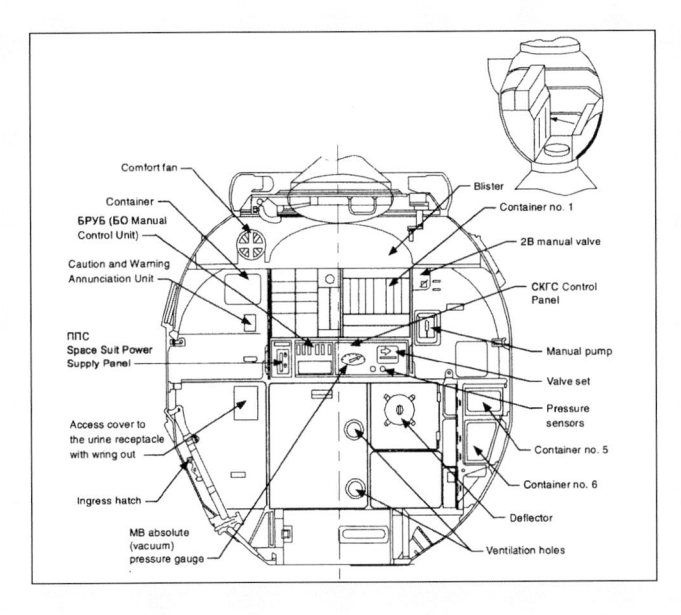

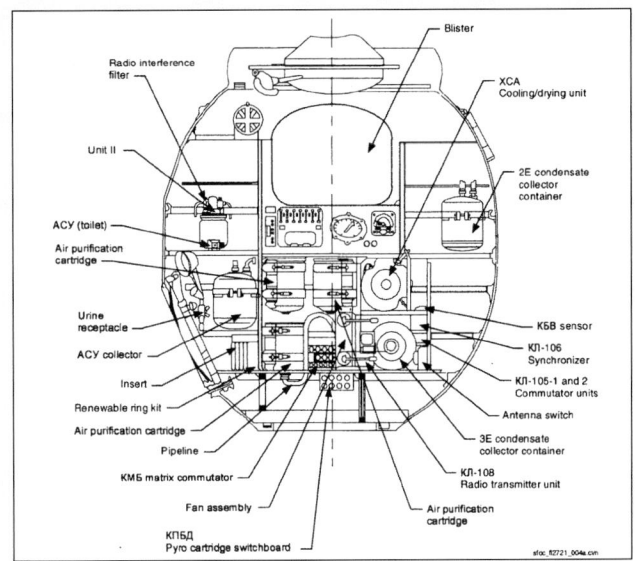

During descent and landing operations, both VHF and SW frequencies are used in a unique direction-finding mode to facilitate converging recovery forces. If the descent module lands in the prescribed area, the VHF signal is used to determine the precise location of the spacecraft. If it lands out of area the SW signal is used first to locate the spacecraft, after which VHF communications begin. Because the main purpose of the SW is as a location finder, the primary means of location is VHF, although during descent both VHF and SW are operated in simplex mode.

The communication system also supports an intercom capability between the crew with all radio conversation recorded on the Gnom-S tape recorder, which is supported by four cassettes on board. One cassette is used for all flight phases up to the completion of rendezvous operations, the second for docking operations, the third for covering activities from spacecraft undocking to landing and a fourth is carried as a spare. In addition there is a P-855A-1 emergency communications radio set for post-landing recovery. Two short-wave transmitters and two receivers are installed within the communication system with a third received for ground communications after landing.

During orbital flight, VHF communication is conducted using the A5M-272 antenna on the spacecraft, but the SW antenna is located on the extremities of the solar array wings and these can only be used after deployment. Moreover, in the descent phase the VHF link is made via the ABM-273 slot antenna situated on the hatch door while the SW link can only be made after the parachute is deployed. From parachute deployment until touchdown, communications via the SW transmitter are effected through the antenna situated on the ropes holding the canopy.

As a functional part of the communications section, the Klest television system is configured to provide live transmission of the crew image from the pre-launch activities to the orbital phases of the flight, including docking. It is also used to televise and monitor graphics and data on the Strelka-VKU sensor equipment, the Simvol equipment and descent parameters on instrumented read-outs in the descent module.

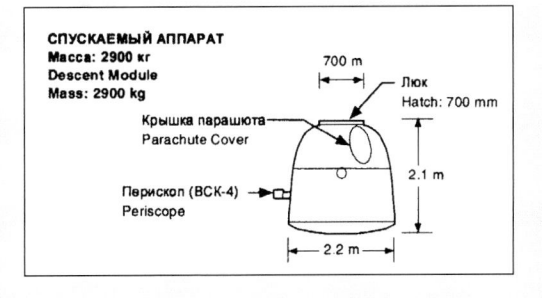

ABOVE LEFT The other half of the orbital module with the covering panel attached, the half of the module which the Russians refer to as the 'sideboard' because it contains a wide range of cabinets and attachments. *(Energia)*

ABOVE The same view of the 'sideboard' with the front-facing panel removed to show equipment, as indicated. *(Energia)*

LEFT External apertures penetrate the mould line for the parachute container, the periscope and the window. *(Energia)*

THE SOYUZ SPACECRAFT

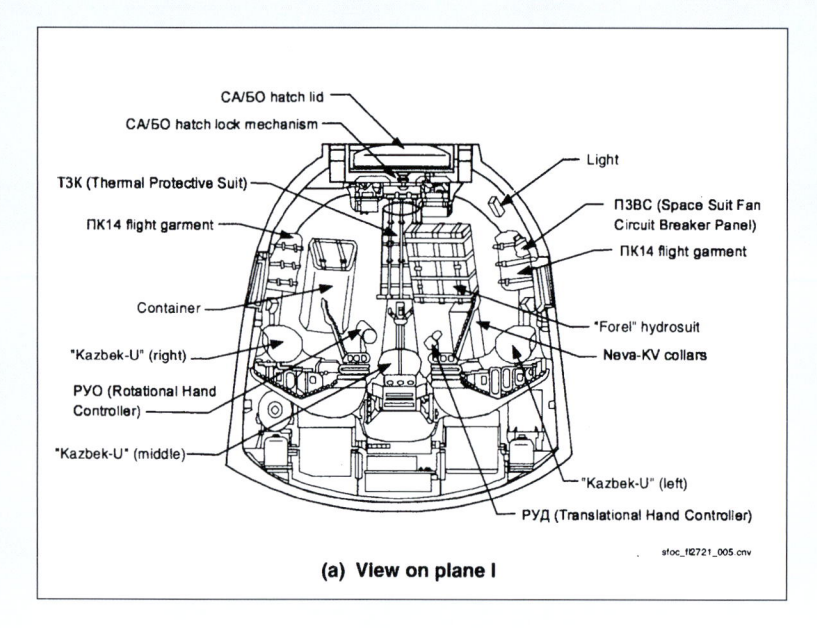

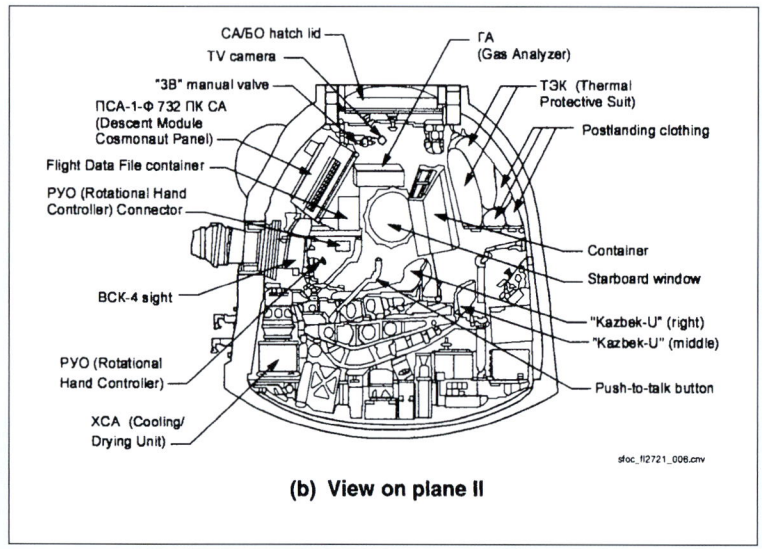

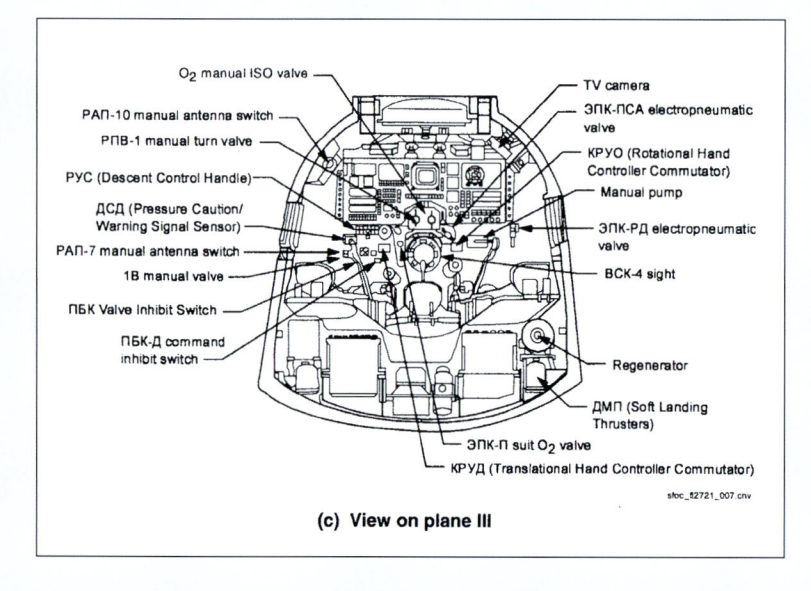

LEFT The interior layout of the descent module seen from the instrument panel, looking in the direction of the cosmonauts' heads, shows the suits container, rotation (attitude control) and translation (movement up, down, forward, backward, up and down) hand controllers. *(Energia)*

CENTRE Looking toward the starboard side from the left flight engineer's seat, with the instrument panel to the left, the side window and below that the periscope, control systems dominate the interior. The cooling/drying unit is part of the environmental control system. *(Energia)*

Cameras are located on the outside of the orbital module with the optical alignment with the spacecraft X (longitudinal) axis, a second camera on the upper starboard side of the descent module and a third camera on the port side. All cameras provide a black-and-white image on a 4:3 frame format and at a 625 line scan frequency on a 25Hz frame rate. The Klest was a Secam analogue system interfacing with the Kvant telemetry unit and transmitting at 463MHz.

The radio telemetry system is designed to sample, memorise and convert sensor data for transmission to the ground stations. The direct measurement/transmission mode is capable of handling 25,600 measurements per second and can record data on three time modes: 32min, 130min and 960min. Two redundant telemetry antennae are located on the aft section of the instrument module skirt working in conjunction with the Kvant-B system transmitter. The telemetry transmission system has 64 channels. In addition, there is a multichannel recorder for recording spacecraft system operations recorded on armoured magnetic tape with a sampling frequency of 31.25Hz. This system

LEFT The view the crew gets looking directly across toward the main instrument and display console. Usually the commander is seated in the middle, flanked by two flight engineer positions. Note the landing thrusters, four of which are housed below the pressurised module and covered during re-entry by the base heat shield, which is jettisoned in the lower atmosphere. *(Energia)*

RIGHT Looking across from the starboard side, the port window allows a limited view during orbital operations. *(Energia)*

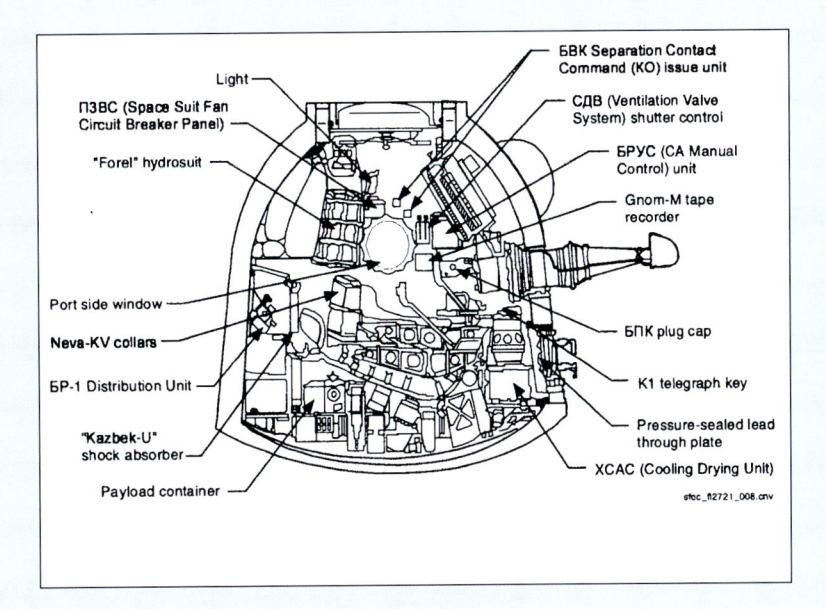

has a capacity of recording 6,000 data points per second with a continuous operating time of 76 minutes.

Command and control systems

Over time, the Soyuz spacecraft has evolved not only in accordance with emerging technologies but also as a necessary part of an integrated system involving space stations. With Soyuz initially seen as an independent spacecraft serving lunar and Earth-orbiting roles, the development of the vehicle into a ferry for plying to and from low Earth orbit brought unique pressure to integrate command and control functions. In this way both Soyuz and the increasingly sophisticated and capable space stations it was adapted to serve displayed common sets of hardware and operating software. Computers came to play an increasingly dominant role in operations.

The first truly 'modern' Soyuz, the TM series incorporated three operationally functional systems. The onboard computer system (OCS) was responsible for taking data from other systems and from onboard and off-board sensors and calculating solutions to attitude control and orientation requirements. The onboard digital computer complex (ODCC) carried the software, which was only infrequently updated, for controlling attitude changes. It was also a mass memory device for storing pre-programmed commands, algorithms and for integrating the several different computers. The onboard complex control system (OCCS) is the processing 'mind' of the system's brain, making decisions, controlling data in and out and commanding routing of data, power switching and network hardware. Essentially, for practical purposes, the functional role of the OCS lies within the operating box of the ODCC and can be considered a supplementary subset in that architecture. It contains the Argon computer and its machine interfaces and the fixed memory. Both the ODCC (incorporating the OCS) and the telemetry management system (TMS) are integrated, with information flowing from the

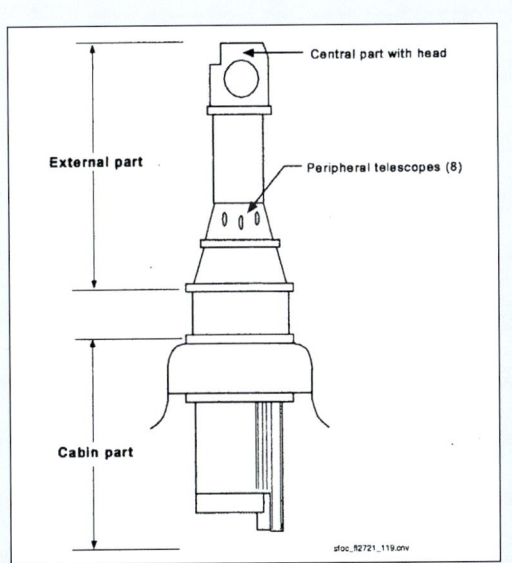

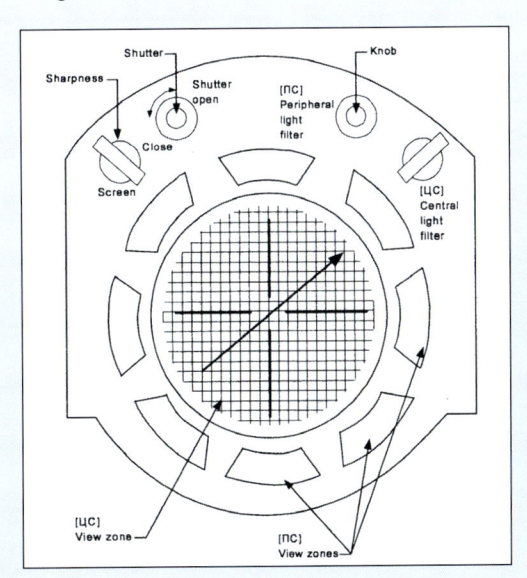

BELOW LEFT The periscope is redundant with the automatic Kurs docking system, but manual use of the TORU remote control docking system requires visual identification of the target. *(Energia)*

LEFT The display panel for the periscope is equipped with focusing and light contrast controls as well as electrical operating switches for adjusting the forward view. The image is displayed on the screen, which has a rotating grid with a line scan system at 1° intervals. *(Energia)*

OCCS to the ODCC and from there to the TMS, which also has a two-way link to the ODCC.

The main computer system initially installed in the early TM series was a derivative model of the Argon generation. It is identified as the Argon-16 from the 16 bits of word or 32 double-word lengths. The Argon series was derived from a particularly robust design from the Kalmykov SAM factory. Designed in the early 1970s by N.N. Solovyov, A.T. Yeryomin and G.D. Monakhov, and developed by F.S. Vlasov, the Argon-16 is indicative of a typical Russian approach to computer design and application. It was developed specifically for the Mir and Almaz space stations and for the Soyuz TM series and the Progress supply vehicles, with production beginning in 1974.

Argon-16 is defined as a synchronous computer with triple redundancy comprising three computers with separate data channels and interfaces to the spacecraft control systems. It has a 3 x 2 KB RAM capacity with 3 x 16 KB ROM. It has a maximum data rate of 80,000kB/sec and contains a computing function in addition to the RAM and ROM modules it has an exchange unit and an interface to the relevant systems, with circuit boards assembled in a book format with rubber spine and forced ventilation cooling.

With a weight of 154lb (70kg) it has a power consumption of 280W and a mean-time-between-failure (MTBF) of 10,000 hours. The Argon-16 is particularly robust and has a solid record for reliability and consistency with only a few recorded failures, notably during an attempted docking with the Salyut space station when it executed the start of an automated docking and then failed, shutting down the operation. The crew was able to recover in the manual mode.

Designed by the Research Institute NII Argon, formerly the Research Institute of Computer Engineering of the Soviet Ministry of Radio Engineering, it was the latest in a long line of evolving computer systems in the USSR, with earlier and contemporary models being developed for the Soviet ballistic missile programmes as well as special defence requirements, including hardening and survivability under an electro-magnetic impulse caused by nuclear detonation.

Developed as a contemporary with the Shuttle AP-101 General Purpose (digital) Computer in the United States, the Argon-16 is a classic example of how Soviet computer design engineers differed from their US counterparts. A Western preoccupation with digital computers belies the clear advantages of analogue systems, which frequently are more reliable and more robust.

While considering why the Russians chose to retain analogue for so long, it is worthy of note to reflect that in the West the science of computers and computing was considered a branch of mathematics rather than an engineering application. Thus in the United

BELOW The upper (forward) hatch that allows access between the orbital and descent modules forms a tight seal before the OM is jettisoned. *(Energia)*

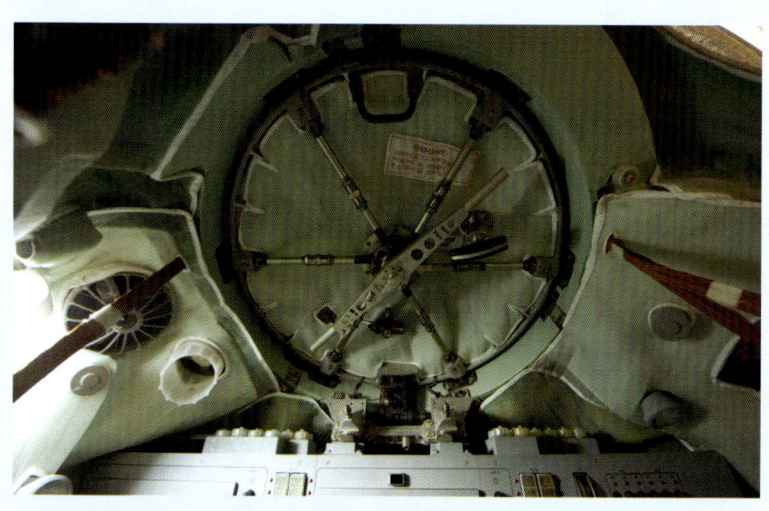

BELOW The hatch can be operated manually using a removable handle. It consists of a series of rotary levers with rollers on the end linked by tie rods. It can also be operated automatically. *(NASA)*

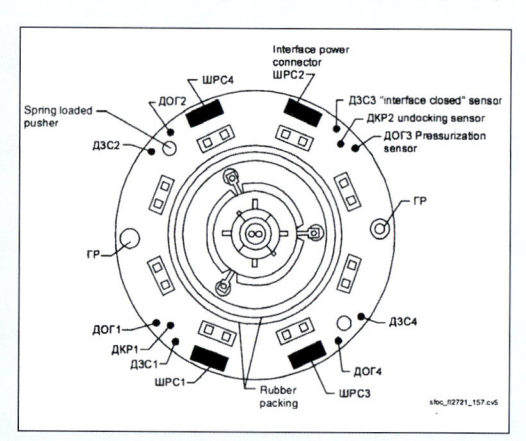

RIGHT The interior arrangement of the seats presents an asymmetry due to the circular cross-section of the descent module. *(NASA)*

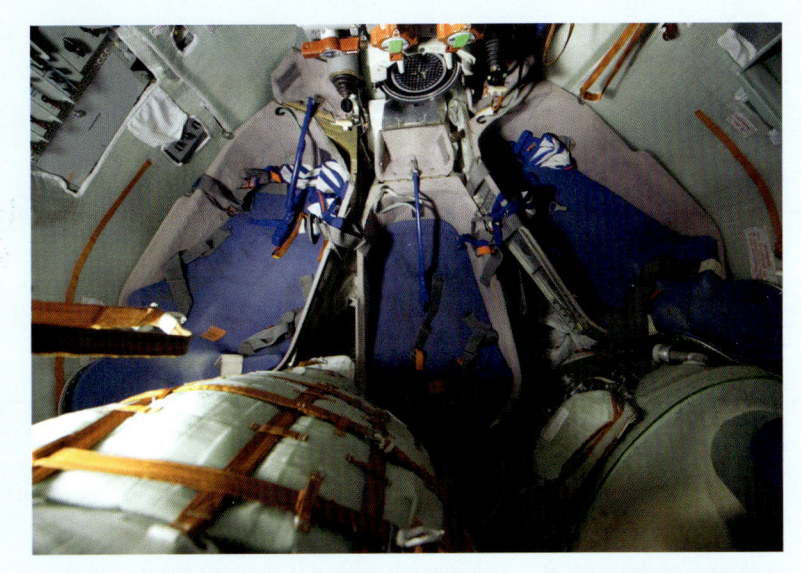

States, paradoxically the home of 'systems engineering' since the 1950s, computer sciences were just that – science. The result was a lack of intuitive engineering skill in both the hardware and the software resulting from digital computer technology.

Even today, computer students are reluctant to apply principles of systems engineering within software programming, which is why it is frequently the software that fails the hardware, and gives computers a bad reputation! In the USSR, those principles of systems engineering were retained far longer than they survived in the West. It is for this reason that Russian computing systems were considered to be inferior and less capable when in fact, by being less sophisticated for a given application, they only appeared lacking in sophistication. This design and engineering philosophy is endemic through the Soyuz system, no less so than in the control electronics segment.

In the design flow for the Soyuz TM, the ODCC carried most of the algorithms for rendezvous and docking operations, by far the most demanding and complex of automated operations conducted on a standard mission. This is the interface with the Mission Control system thrusters which are operated on commands received from the OCS, which itself responds to sensors at an interface with the ODCC. The ODCC also contains the software patches for inputting to the RAM in the OCS, and because it is prioritised to look first at the RAM for instructions, software patches are loaded there first.

Relatively new in the mid-1970s, the universal computer programmer language C was used for the Argon-16. Developed by Dennis Ritchie, it would quickly become one of the most universally adopted languages in the computer world and was used for the majority of computer architectures around the globe – including the Soviet Union. Its value lay in its simplicity and low-level capability, which made it an ideal base for cross-platform programming, a particularly useful base from which to build the software for the Argon-16.

With the introduction of the TMA-M series in 2010, the Argon-16 was replaced with the TsVM-101, a digital computer capable of many more functions and with less weight and

BELOW LEFT The display and controls panel is located on the side wall and switches are accessed by the commander using an extended rod to depress switches. *(NASA)*

BELOW The greatest area of upgrade has given the crews better computer display and controls for both access and flexibility of the digital equipment now installed in the descent module. *(NASA)*

77

THE SOYUZ SPACECRAFT

ABOVE The Globus panel that evolved from the equipment designed for the early Voskhod spacecraft was developed for early versions of Soyuz with an electro-mechanical position locator over a spherical globe of the Earth. *(David Baker)*

ABOVE The Globus navigation instrument is a direct development of the simplified panel installed in the first Vostok spacecraft as indicated by this preserved example. *(David Baker)*

reduced power demands. The new computer has a mass of approximately 57lb (26kg) with a power consumption of 80W and is integrated with the guidance, navigation and control system. The RAM capacity is increased to 2,000kB with a processing rate of eight million operations per second. Operational life is extended to 35,000 hours. Other improvements to the guidance and navigation system have reduced power consumption from 402W to 105W, greatly reducing the power requirement from the solar arrays and battery equipment.

Crew access to the ODCC was through a set of input keys on the main display console. Called the Neptune panel, it was equipped to receive commands as well as for monitoring the various spacecraft systems, for using the TV systems and for the caution and warning systems. Physical access for attitude and orientation commands to the ODCC was through two hand controllers. The attitude control handle (ACH) allowed manual authority over the spacecraft pitch, roll and yaw axes for aligning axial pointing modes. For manoeuvring the spacecraft back, forward, left, right, up or down, and for orbital trajectory changes, the crew had the translational motion control handle (TMCH).

For visual identification of relative location with respect to the ground track of the spacecraft's orbit, and for visual cues as to the specific orientation of the vehicle, the Globus device provided a view of the Earth below as part of the guidance, navigation and control function. This was the Russian equivalent of the flight director attitude indicator (FDAI) – or '8-ball' – in NASA space vehicles from the Gemini and Apollo spacecraft of the 1960s to the Shuttle of the 1980s and beyond. As part of the Neptune panel, it shares space with the communications equipment and headset controls, a caution and warning display, electrical displays (voltage and current) and a keyboard for command inputs.

Not for semantic reasons do the Russians sometimes refer to the OCCS as the onboard complex control and management system (OCCMS), but rather because the control system is primarily responsible as a power switching, command processing and data-routing command unit with a limited degree of crew interface in flight. While US spacecraft have distinct functionality within separate independent systems, converging at the primary processing centre, or computer, Soyuz spacecraft have a more open architecture. The OCCS is responsible not only for controlling powered-up systems or subsystems, but also

ABOVE The detailed engineering in the Globus panel is akin to the perfection expected of a Swiss watchmaker and demonstrates a unique form of electro-mechanical computer/prediction system that was applied to perfection in the USSR. *(David Baker)*

LEFT Variable resistors, boards and diodes populate the workings of the Globus panel as built into Voskhod. It was from these that all early Soyuz position and navigation information displays evolved. *(David Baker)*

for switching electrical power between these elements, effectively turning them on or off as required by the central switching function of the OCCS.

Herein lay another departure from US practice, in that the degree of computer control extends to execution of tasks, switching systems and subsystems on and off as required to conduct what is a fully automated set of activities. In US spacecraft, the philosophy has always been that the crew conduct the task, manually switching on or off the various systems and subsystems required to accomplish that. It was a difference in control approach that had dictated the way the Soyuz systems had been set out and was one of the more challenging aspects of integrating US and Russian crew members, especially those who had grown up with the Apollo and Shuttle systems' philosophy.

The OCCS command processing unit is central to performing programme logic and for authorising priority selection among the systems. A matrix decoder receives 23-bit signal commands from the ground and sends the appropriate response to the command radio link for downlink. The timing device supplies timing information to the command processing unit with additional functionality for time references and the storage and transmission of timing signals. In addition, the indication transmitter sends compressed status and signal information to the command processing unit. The OCSS also routes raw data to the telemetry management system, where it is processed into transmission packets.

Because the selection of hardwired analogue computers had been driven by the philosophy of automation throughout Soyuz operations, their very existence optimised this approach to management of the mission. As a measure of how far this penetrated into the design approach, incoming ground commands had priority over manual commands on board the spacecraft, a legacy of the Korolev days when it was believed that cosmonauts should be passengers and forego any piloting role they had been used to in a former life.

As a footnote to the major systems changes affected as a result of mission-creep, when Soyuz became a taxi and not merely an

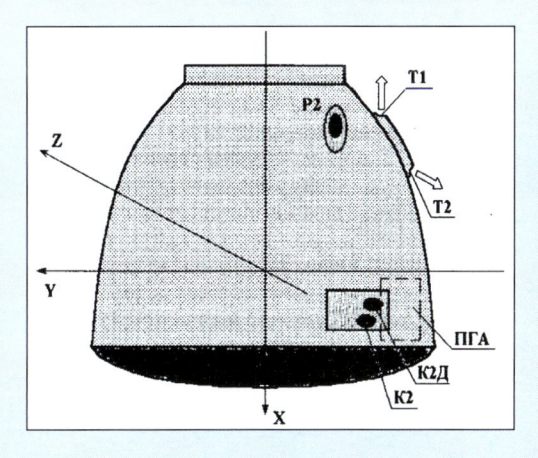

LEFT Thruster locations on the descent module for attitude control during re-entry which, although desirable for controlling a guided or a ballistic trajectory, are not essential for the survival of the crew. *(Energia)*

79

THE SOYUZ SPACECRAFT

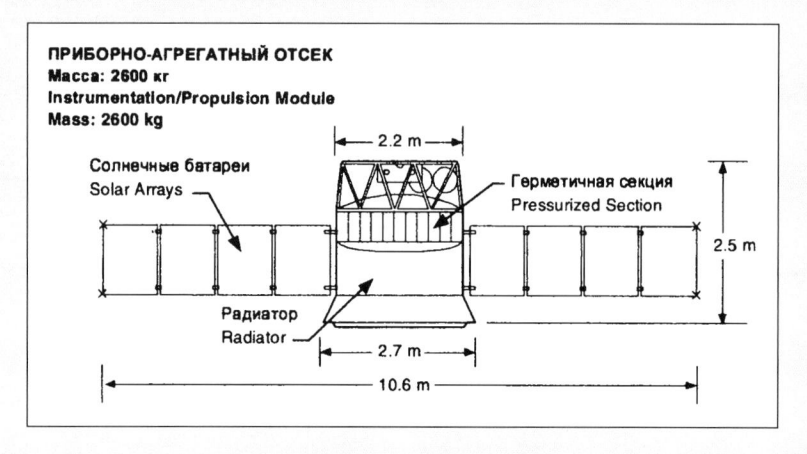

ПРИБОРНО-АГРЕГАТНЫЙ ОТСЕК
Масса: 2600 кг
Instrumentation/Propulsion Module
Mass: 2600 kg

Солнечные батареи / Solar Arrays
Герметичная секция / Pressurized Section
Радиатор / Radiator

2.2 m
2.5 m
2.7 m
10.6 m

Key:
1. ДПО (Approach and Attitude Control Thrusters)
2. БК-3 Oxygen Tank (4 pcs)
3. РКО (Orbit Radio Tracking) Antenna
4. Truss
5. Pyro lock
6. [AC-17] Antenna (2 pcs)
7. Spring pusher
8. Heat exchange manifold
9. Ring
10. Electrical disconnects

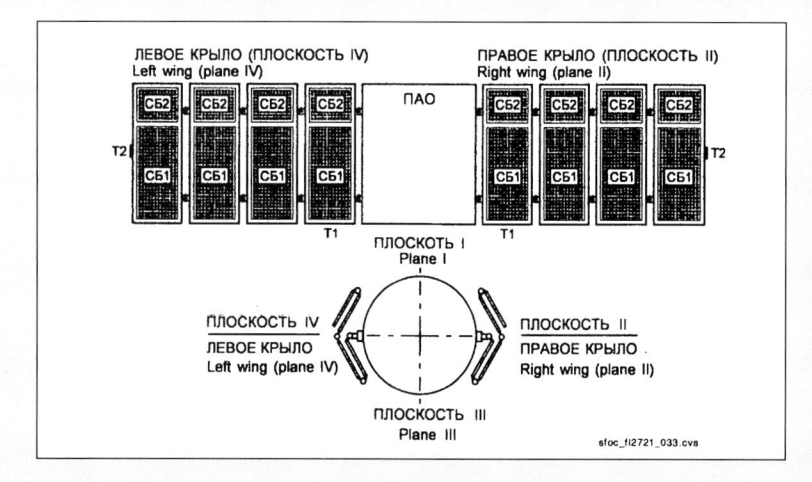

ЛЕВОЕ КРЫЛО (ПЛОСКОСТЬ IV) / Left wing (plane IV)
ПРАВОЕ КРЫЛО (ПЛОСКОСТЬ II) / Right wing (plane II)
ПАО
СБ2 / СБ1
T1, T2
ПЛОСКОТЬ I / Plane I
ПЛОСКОСТЬ IV / ЛЕВОЕ КРЫЛО / Left wing (plane IV)
ПЛОСКОСТЬ II / ПРАВОЕ КРЫЛО / Right wing (plane II)
ПЛОСКОСТЬ III / Plane III

LEFT The arrangement of the instrument/propulsion module, divided into pressurised and unpressurised compartments, contains all essential systems, thrusters and retro-rockets for control of the spacecraft's life support, manoeuvring, attitude control, communications and electrical power. *(Energia)*

CENTRE The adapter section consists of a truss assembly connecting the instrument module with the descent module and provides fixtures for four oxygen tanks, electrical umbilicals and associated harnesses. *(Energia)*

autonomous spacecraft, the basic functionality of the systems architecture and computer, as well as software design, was followed with the Mir space station. In it, two Argon-16 computers were installed, one more capable than the unit installed in Soyuz, and with higher capacity for a wider range of operations. But the philosophy of fully automated operation, with manual override, was the fundamental starting point for the design of Russian space stations.

Lessons learned through use of the older Soyuz, and earlier Salyut stations, resulted in changes to the computers aboard Mir. The older Delta navigation computer on the Salyut 7, which required continuous updates from the ground, was replaced with a system designated EVM which was capable, in theory, of controlling the complex without any human presence or intervention. The Delta system had been so limited that its operation dominated control of the Salyut complex with or without a crew, taking time that would otherwise be spent on scientific research and operating experiments.

Critics have claimed that this almost

LEFT The solar arrays are folded against the side of the instrument module until separating from the second stage of the launch vehicle. With a surface area of 108ft^2 (10m^2), the arrays are capable of providing at least 26A at 34V when they are orientated to within +/-10° of the Sun. Orientation is achieved by moving the spacecraft in its y (yaw) axis. The arrays are connected in series, with a switching capability to deactivate the left wing if necessary. *(Energia)*

RIGHT The power supply and distribution buses with their respective switching circuits are separated into separate circuits with remote and manual circuit breakers. Power-switching elements in the instrument module consist of remote control breakers, with both descent and orbital modules equipped with each. *(Energia)*

1. ПАО	- Instrumentation/Propulsion Module	8. БСАС	- Power Supply Automation Unit, Descent Module
2. СЭП	- Electrical Power System	9. КАБЕЛЬ-МАЧТА	- Cable Conduit
3. ПК, КР-Д2 "ОБЪЕДИНЁННОЕ ПИТАНИЕ" ОТКЛ ("АВТОНОМНОЕ ПИТАНИЕ") -	Crew console one-time command Д-2 "INTEGRATED POWER" OFF ("AUTONOMOUS SUPPLY")	10. ОВК12/ОВК 13 "ОТСТРЕЛ БО"	- Critical Command ОВК12/13 Orbital Module Jettison
		11. ОК53 "РАЗДЕЛЕНИЕ"	- Gen CMD ОК53 "SEPARATION"
		12. БСА	- Descent Module Battery
4. ПК, КР-Д2 "ОБЪЕДИНЁННОЕ ПИТАНИЕ" -	Crew console one-time command Д-1 "INTEGRATED POWER"	13. ГЕРМОПЛАТА	- Pressurized Circuit Board
		14. ПИРОНОЖ	- Pyrotechnics (used to cut the cable conduit before БО separation)
5. БКБП	- Onboard Power Supply Switching Unit	15. БО	- Orbital Module
6. БСАП	- Power Supply Automation Unit, Instrumentation/Propulsion Module	16. БСАБ	- Power Supply Automation Unit, Orbital Module
		17. ШРС1	- Interface Power Connector 1
7. СА	- Descent Module	18. ШРС2	- Interface Power Connector 2

overbearing reliance on automated systems is a reflection of an authoritarian Soviet system from which the spacecraft design philosophy emerged. A more pragmatic interpretation of this philosophy has it that all space stations should be autonomous, freeing the crew to do productive work rather than assigning time to maintaining systems and subsystems in a routine working order.

Reality lies somewhere in between, and for a complex as large and operationally sophisticated as the International Space Station many functions can be transferred to automated control, but the time allowed for maintenance and repair is necessarily

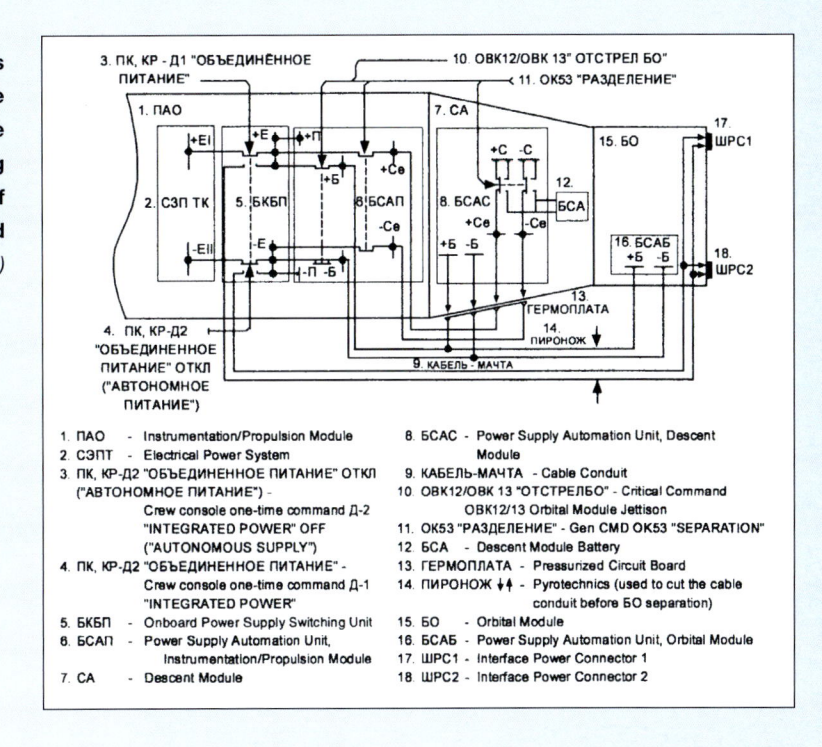

BELOW The power supply system provides a feed directly from the Soyuz to the International Space Station and allows power to be directed from the ISS Russian sector to the Soyuz. *(Energia)*

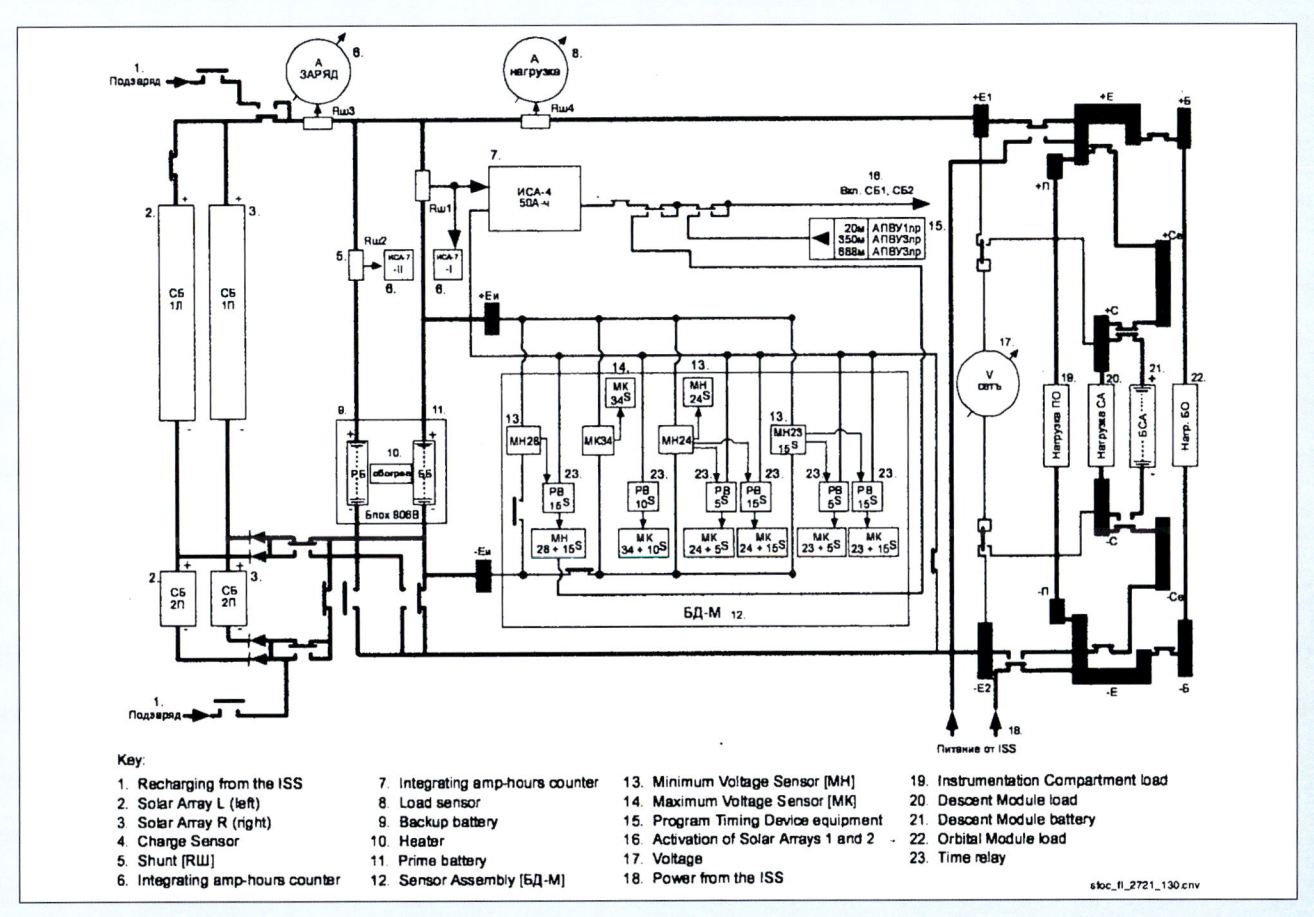

Key:
1. Recharging from the ISS
2. Solar Array L (left)
3. Solar Array R (right)
4. Charge Sensor
5. Shunt [РШ]
6. Integrating amp-hours counter
7. Integrating amp-hours counter
8. Load sensor
9. Backup battery
10. Heater
11. Prime battery
12. Sensor Assembly [БД-М]
13. Minimum Voltage Sensor [МН]
14. Maximum Voltage Sensor [МК]
15. Program Timing Device equipment
16. Activation of Solar Arrays 1 and 2
17. Voltage
18. Power from the ISS
19. Instrumentation Compartment load
20. Descent Module load
21. Descent Module battery
22. Orbital Module load
23. Time relay

81

THE SOYUZ SPACECRAFT

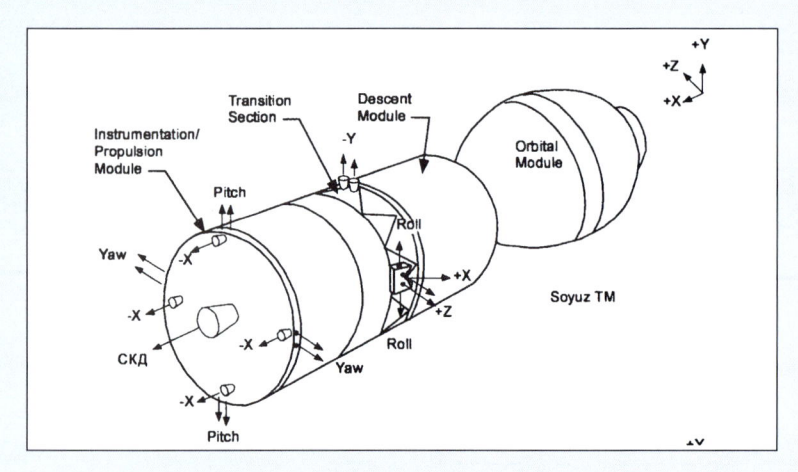

ABOVE Attitude, orientation and manoeuvring thrusters are located on the propulsion module and at selected locations on the transition truss for control about roll, pitch and yaw axes. *(Energia)*

ABOVE A schematic showing the layout and distribution of the numbered thrusters for the attitude and orientation system. *(Energia)*

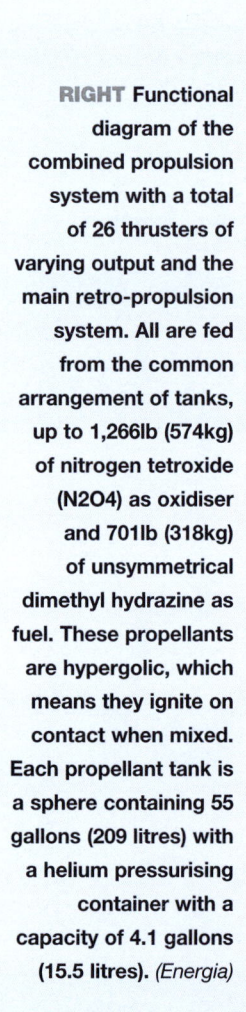

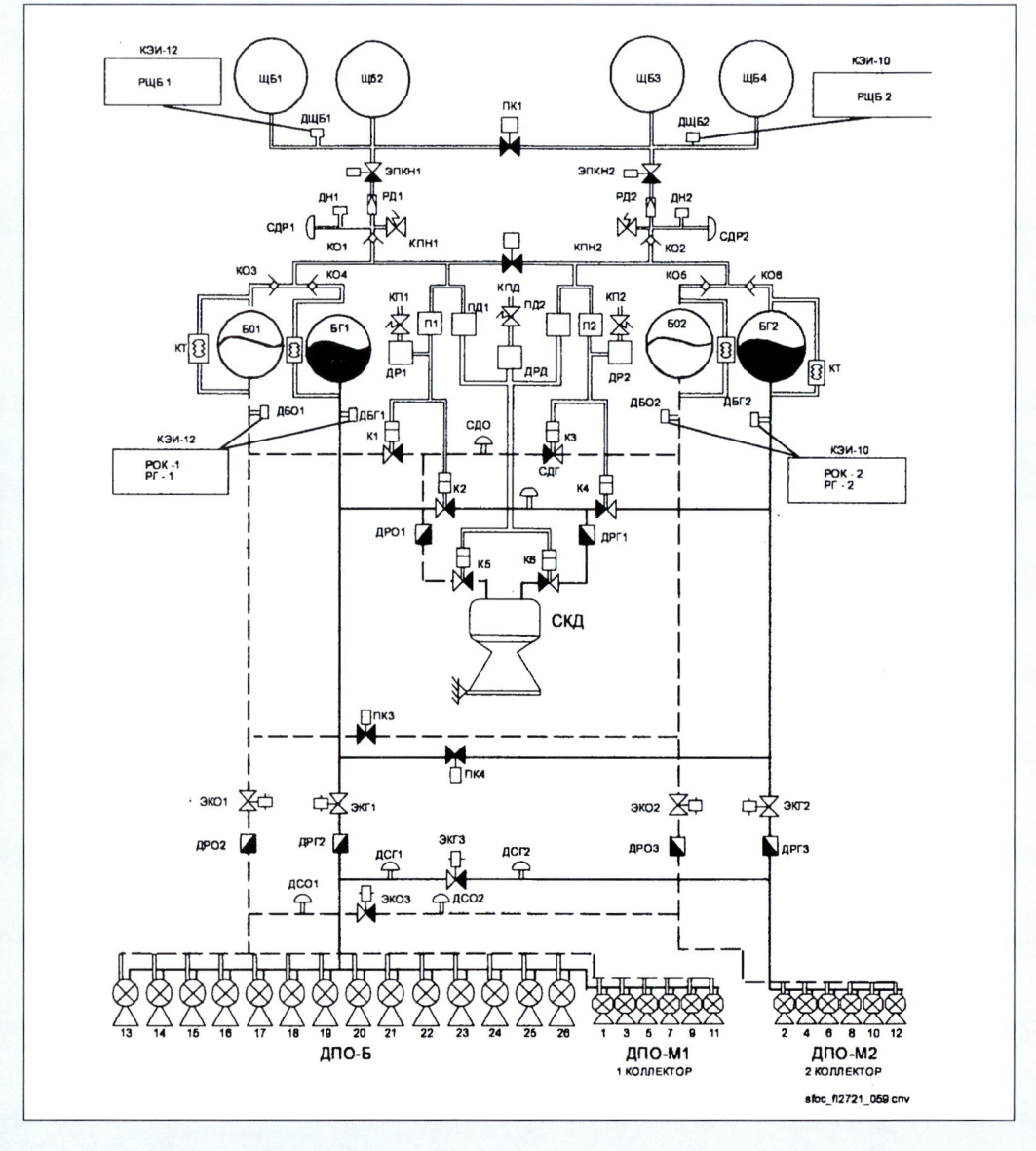

RIGHT Functional diagram of the combined propulsion system with a total of 26 thrusters of varying output and the main retro-propulsion system. All are fed from the common arrangement of tanks, up to 1,266lb (574kg) of nitrogen tetroxide (N2O4) as oxidiser and 701lb (318kg) of unsymmetrical dimethyl hydrazine as fuel. These propellants are hypergolic, which means they ignite on contact when mixed. Each propellant tank is a sphere containing 55 gallons (209 litres) with a helium pressurising container with a capacity of 4.1 gallons (15.5 litres). *(Energia)*

82
SOYUZ MANUAL

RIGHT The Soyuz propulsion pressurisation system, with helium for expelling the propellants under pressure to the respective thrusters and engine. *(Energia)*

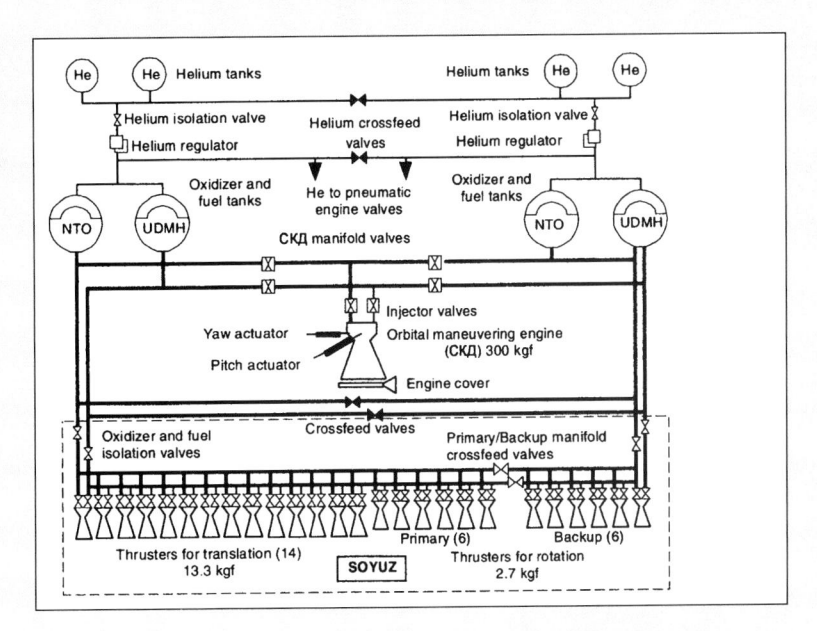

increased with the total number of systems and engineering components. For a spacecraft such as the Soyuz ferry vehicle a significant level of automation is both time-saving and prone to fewer incidents of human error.

The operational protocol for the telemetry system has evolved over the life of Soyuz, and has shifted operational strategies according to the prevailing conditions in the country. Before the collapse of the Soviet Union, Russia had several tracking ships deployed to cover Atlantic and Pacific Ocean regions where there were no geographic sites for land-based tracking stations. Politically isolated from many other developed countries, Russia did not have the luxury afforded to the United States by the cooperation of many friendly countries where such facilities could be set up for communicating with US spacecraft.

After the disintegration of the USSR, Russia sold off the ships that had carried out these tracking and communications functions, and orbital coverage was significantly reduced as a result. Moreover, Russia did not invest in the geostationary data relay satellites such as those launched by the United States for complete coverage of Shuttle flights and now the International Space Station. Only the launch and ascent phases were covered without pause, data flowing continuously to the Mission Control Centre in Moscow.

The BR-9CU-3 Soyuz TM telemetry system is located in the Instrument Module on an FM downlink capability transmitting at

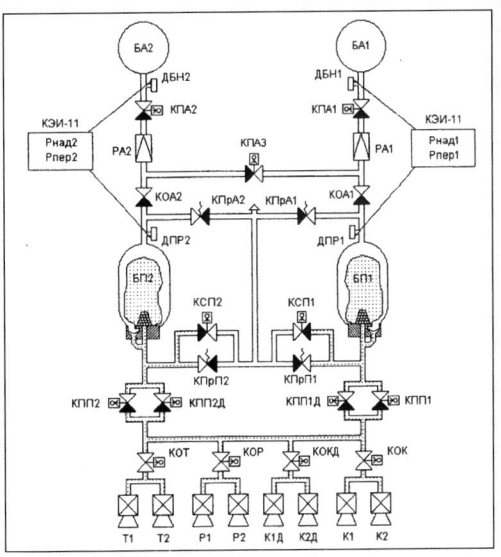

LEFT The propulsion system for the eight thrusters in the descent module is completely independent of the system used on orbit and contains its own monopropellant tanks and supply system. *(Energia)*

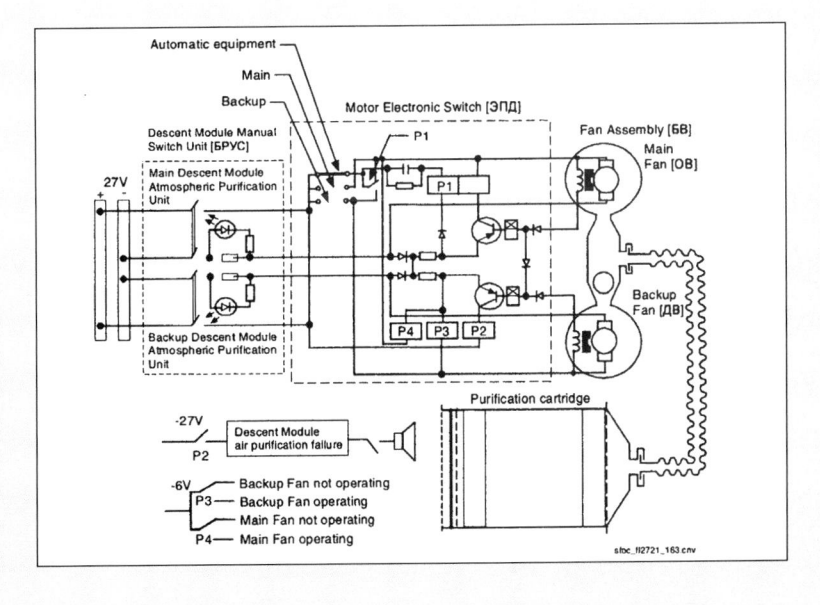

RIGHT Known as the atmosphere purification unit, the removal of carbon dioxide exhaled by the crew is conducted through a unit with absorbent scrubbers carried in the descent module. One purification cartridge is in the DM with other filters stored in the orbital module. The carbon dioxide level is kept below 2–4mm Hg, with each cartridge good for about 60 man-hours (20 hours for a crew of three). *(Energia)*

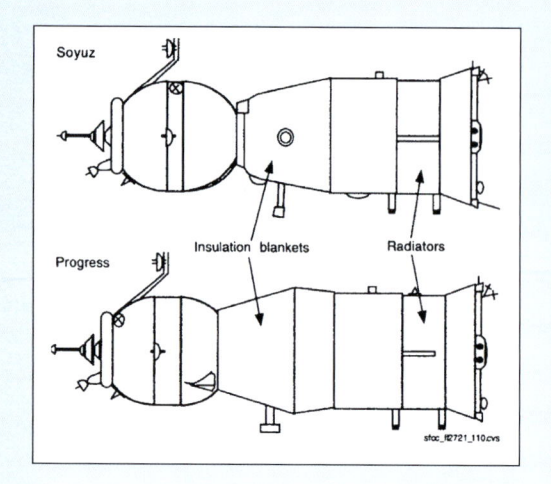

LEFT A passive thermal control system is employed for Soyuz and Progress vehicles to inhibit thermal transfer from inside or outside the vehicle. The insulation consists of an aluminium-coated polyethylene teraphalate film separated by layers of glass fibre. It is highly reflective, minimising heating of the vehicle under high Sun conditions and preventing loss of heat during eclipse periods. *(Energia)*

BELOW The active thermal control system (CTP) maintains an air temperature of 18–25°C in the habitable compartments and 0–40°C in the instrument section, with a relative humidity of 30% to 75%. The CTP consists of five hydraulically isolated cooling loops including internal and external thermal control loops, a water coolant loop, an intermediate heating loop and a condensate evacuation loop. *(Energia)*

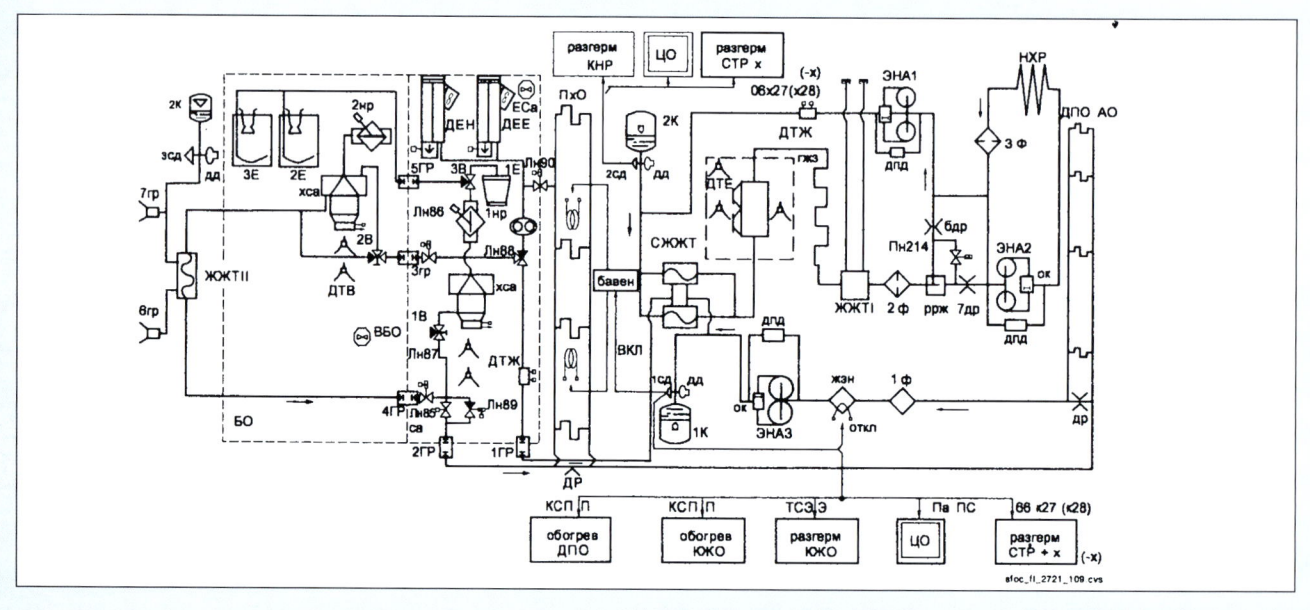

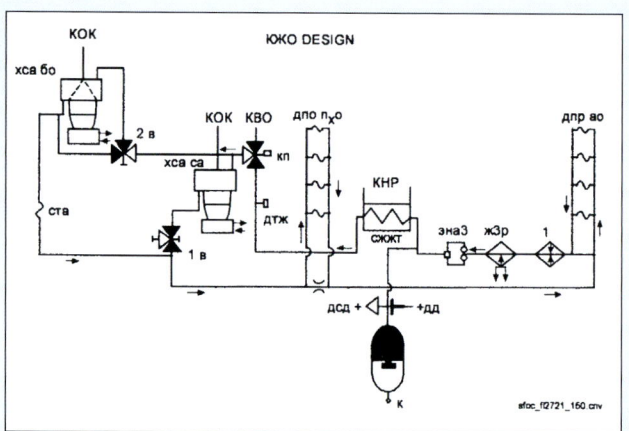

LEFT The internal thermal control loop, which is responsible for maintaining temperature and humidity in the two habitable compartments. *(Energia)*

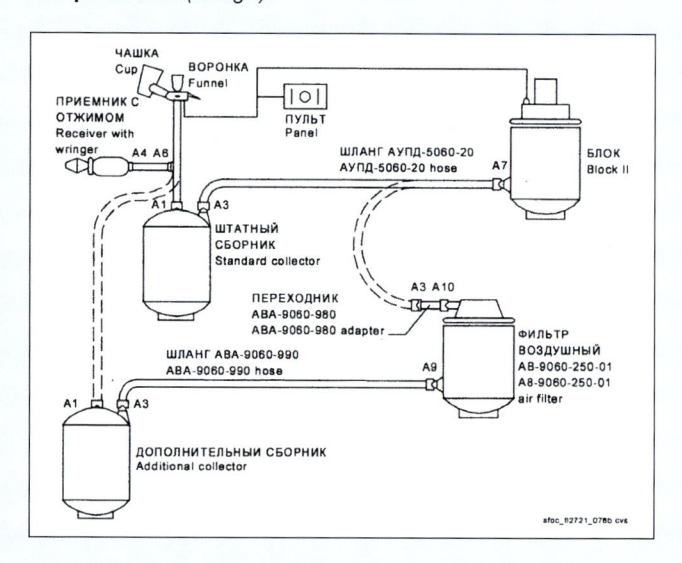

RIGHT The Soyuz toilet is a pneumohydraulic system with an operating life of 12 man/days. It has a collector volume of 2.85 gallons (10.8 litres) and is based on an average urine production level of 0.32 gallons (1.2 litres) per person per day. It consists of quick-disconnect hoses and appropriate receptacles. *(Energia)*

84
SOYUZ MANUAL

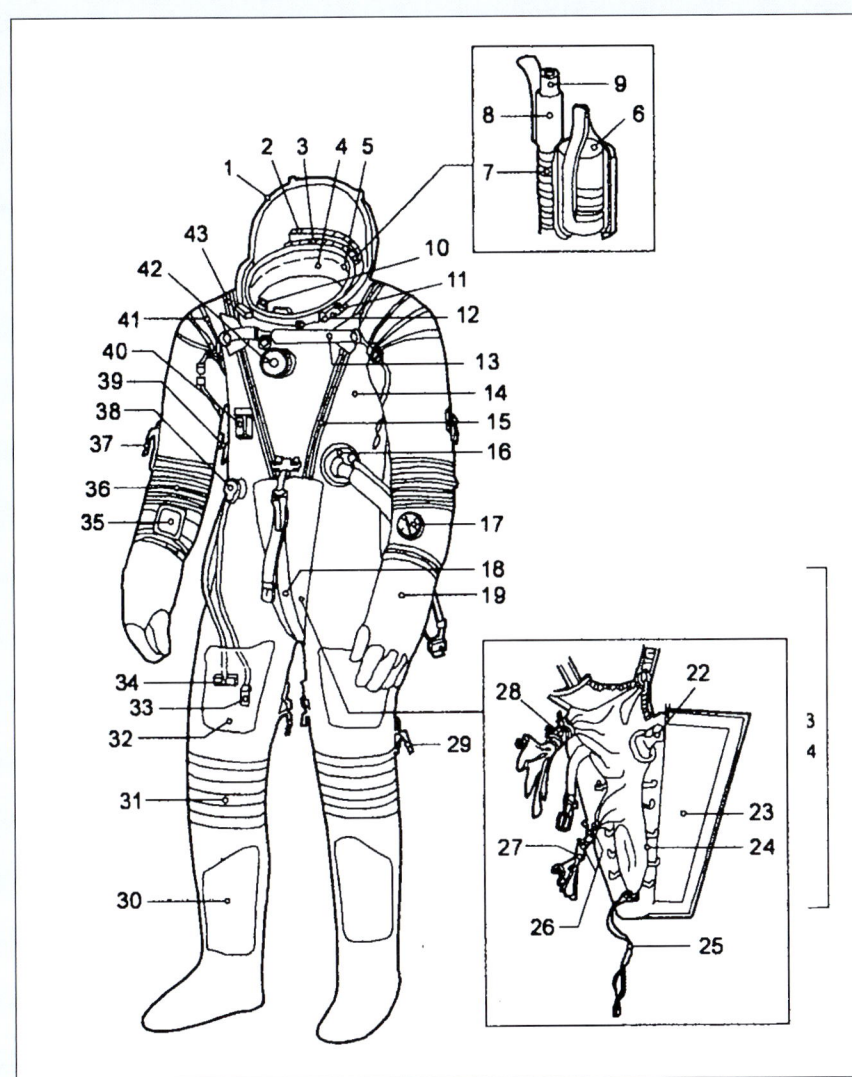

1. Pressure helmet
2. Oxygen manifold
3. Ventilation manifold
4. Protective collar
5. Collar lanyard
6. Inflation valve
7. Ventilation tube
8. Connector
9. Fitting
10. Collar lanyard cotter pin
11. Pressure helmet lock
12. Pressure helmet lock latch
13. Reinforcing straps with buckles snap
14. Pressure-restraint shell
15. Zipper
16. Group inlet
17. УДИС suit pressure gauge
18. Front-draw strap with snap hook
19. ГП-7 pressure glove
20. Oxygen feed hose
21. Ventilation hose
22. Physiological opening with hooks for ?
23. Flap with Velcro fastener
24. Physiological opening with hooks for ?
25. Lace
26. Rubber bands for sealing "appendages"
27. Small "appendage"
28. Large "appendage"
29. Buckle for leg-length adjustment strap
30. Pocket for pressure gloves
31. Knee joint
32. Pocket
33. Electrical connector for communications
34. Medical monitoring electrical connector
35. Mirror
36. Elbow joint
37. Buckle for arm-length adjustment strap
38. Airtight electrical input
39. Buckle for torso-length adjustment strap
40. Pocket for internal electrical connectors, communications, and medical monitoring
41. Shoulder joint
42. РДСП pressure regulator
43. Front opening

166MHz or 922MHz, both at 256kbps. The 38G6 tracking system was co-located with the telemetry system and this transmitted on a two-way link. The Kvant avionics system was installed adjacent to the telemetry and tracking equipment and operated at a rage of frequencies in 700–900MHz range.

While telemetry relay through an orbiting satellite was not possible with the Soyuz TMA vehicle, once docked to Mir the communications facilities of that station could be used in a 'bent-pipe' mode. In this instance, hard-wired signals through the docking interface are sent to the Mir communication and telemetry system and then relayed via a Russian satellite to Moscow. Likewise, command instructions could be uplinked from the ground to Mir and transferred across to the Soyuz spacecraft while it remained docked to the station.

The Sokol spacesuit

The Sokol pressure suit is the individual protective equipment set which is manifested aboard Soyuz as a part of the KCC survival air complex. The suit is a backup to the cabin atmosphere revitalisation system, and as a life preserver in the event of a catastrophic rupturing of the pressurised volume. It works in synergy with the low-pressure sensor system to maintain a pressure upon the crew member of at least 5.8lb/in^2 (300mm Hg). The suit is also designed to serve as a flotation aid in the event of ditching and provides buoyancy by means of a special collar. This collar also serves as a means of lifting the occupant out of the water by helicopter.

The KCC complex of equipment is connected to the spacecraft systems throughout the time the occupant is wearing it,

ABOVE The Sokol pressure suit is designed to preserve life inside the descent module in the event of an unexpected depressurisation and is not suitable for spacewalks, for which an Orlan suit is used. Spacewalks do not take place from Soyuz itself but are made from airlock modules on the International Space Station. *(Energia)*

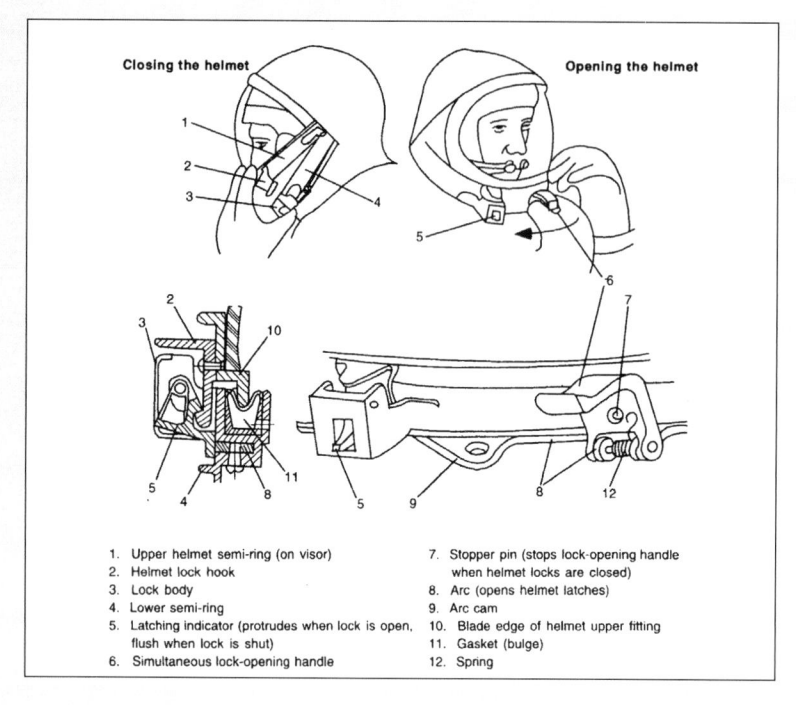

1. Upper helmet semi-ring (on visor)
2. Helmet lock hook
3. Lock body
4. Lower semi-ring
5. Latching indicator (protrudes when lock is open, flush when lock is shut)
6. Simultaneous lock-opening handle
7. Stopper pin (stops lock-opening handle when helmet locks are closed)
8. Arc (opens helmet latches)
9. Arc cam
10. Blade edge of helmet upper fitting
11. Gasket (bulge)
12. Spring

ABOVE The helmet and fittings designed to accommodate a crew member with a manual locking and unlocking latch for pressure integrity. *(Energia)*

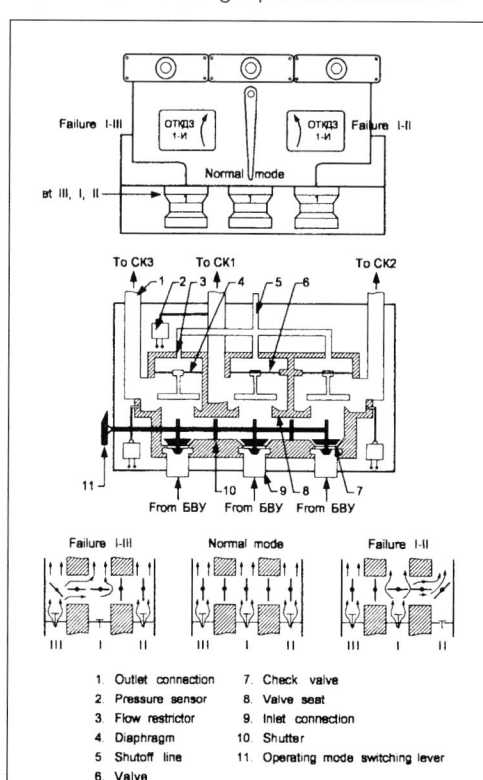

1. Outlet connection
2. Pressure sensor
3. Flow restrictor
4. Diaphragm
5. Shutoff line
6. Valve
7. Check valve
8. Valve seat
9. Inlet connection
10. Shutter
11. Operating mode switching lever

RIGHT The suit distribution unit and associated control assembly for integrating the atmosphere circulation system with the Sokol suit. *(Energia)*

ensuring that automatic cycling of emergency oxygen will be delivered to the suit. As an integral part of the ARS, it is a spacecraft within a spacecraft, and being totally integrated is somewhat different from the system used in the former NASA Shuttle.

In addition to being a pressure bladder to maintain a survivable pressure and oxygen supply, the suit inventory also includes press-tight gloves, boots, underwear, cotton/flax socks and a communication headset. Unlike US suits, there is also provision for in-flight maintenance, including drying aids to vent moisture through evaporation, packets of glass-oil to prevent fogging of the faceplate, glove drying aids, electrical connectors, ear plugs and a special bag to contain the suit when doffed.

The suit itself is more than a survival garment: it is a spacecraft in its own right. It is a soft structure comprising a double-layer garment with built-in soft helmet and removable gloves. There is an opening in front for donning the suit, with a pressure differential indicator on the left cuff, and with a removable mirror on the right cuff to be used when donned for checking various parts of the suit connectors and for examining the helmet locks. The suits are manufactured to a standard size with special adapters for anthropomorphic variations among different crew members. They are designed for comfort in the crouched position in the crew couch in the descent module and this gives the cosmonaut a strangely squat, almost awkward appearance when standing or walking out to the launch vehicle.

The suit is assembled from a tight inner pressure envelope and an external structural shell. The inner envelope is made of rubberised Capron fabric with rubberised knitted fabric in the joints. The pressure envelope also carries an array of elastic pipelines for ventilation and to supply oxygen to the crew member via a port in the helmet. These pipelines also provide ventilation and are found in the leg, sleeve and helmet areas. Special insoles provide ventilation to the feet. A rubberised neck dam prevents the suit leaking when the cosmonaut is in water, but the collar itself is kept folded for all normal modes of flight and landing. A drawstring in the collar allows the cosmonaut to pull it closed around the throat to provide a watertight seal.

The so-called structural shell is a flexible outer covering fabricated from Lavsan fabric with built-in structural bands with clasps in the sleeves and trouser areas for fitting. The occupant dons the suit through the front opening closed with two structural zippers. The frontal opening is reinforced with a transverse structural band that locks the shoulder joint cords, fastened by a spring lock in the waist area. Soft joints in the shoulder, elbow and knee areas afford a certain degree of mobility, with cord connectors helping with movement. Adjustable front tightening bands allow a degree of 'give' when the internal pressure layer is inflated.

The helmet comes integral with the top of the suit, locked when two half-frames are in contact. A mechanical flag indicator reveals incorrect closing. Ventilation and oxygen manifolds are positioned within the helmet for circulating air and for oxygen in the event of a depressurisation. The helmet is fabricated from polycarbonate plastic with an external cover of organic material to protect it from scratches until the crew member is inside the descent module.

The suit gloves comprise an inner pressure retainer and an external structural shell with finger sections fabricated from Capron knitwear covered with rubber. Special bulbous sections are fitted in the finger joints, while the shell itself is made from Lavsan. Pressure gloves are fabricated in three sizes – the only variable elements of the suits – and come with palm half-grip fitments of 3.15in, 3.35in and 3.54in (8, 8.5 and 9cm). Special hygienic cotton gloves are donned before the gloves are fitted.

The suit pressure regulator unit allows crew members to breathe normal air with the helmet closed and when there is no suit ventilation. There are two different pressure modes for each crew member to select for pressurisation. A special pressure differential manometric indicator (PDMI) provides for visual monitoring of the suit pressure and has a range of graduated scales to indicate inflation rate. Electrical equipment within the suit provides for biomedical monitoring from body-placed sensors and for radio communications with the outside world. Bunched connectors link these outlets to ground equipment for validation and testing and to connectors hooking up the suit circuit to the spacecraft itself.

With the occupant plugged in to the spacecraft suit circuit, he or she is protected by an automatic system which ensures that, should pressure in the oxygen/nitrogen cabin atmosphere fall below 7.7lb/in² (400mm Hg), ventilation air to the suit is cut off and oxygen from the stowage tank will flow to the suit. As pressure drops to 5.8lb/in² (300mm Hg), the pressure equalisation valve closes and a pure oxygen environment is maintained at that level within the suit.

Humans can survive a pure oxygen environment at this level for several weeks, but the partial pressure of oxygen at sea level in the Earth's atmosphere is about 3lb/in² (155mm Hg), and cerebral anoxia, with attendant brain damage, can result if pressures fall below this level. Flexible suits cannot withstand a differential of sea-level atmospheric pressure on the inside and a vacuum on the outside, so it is

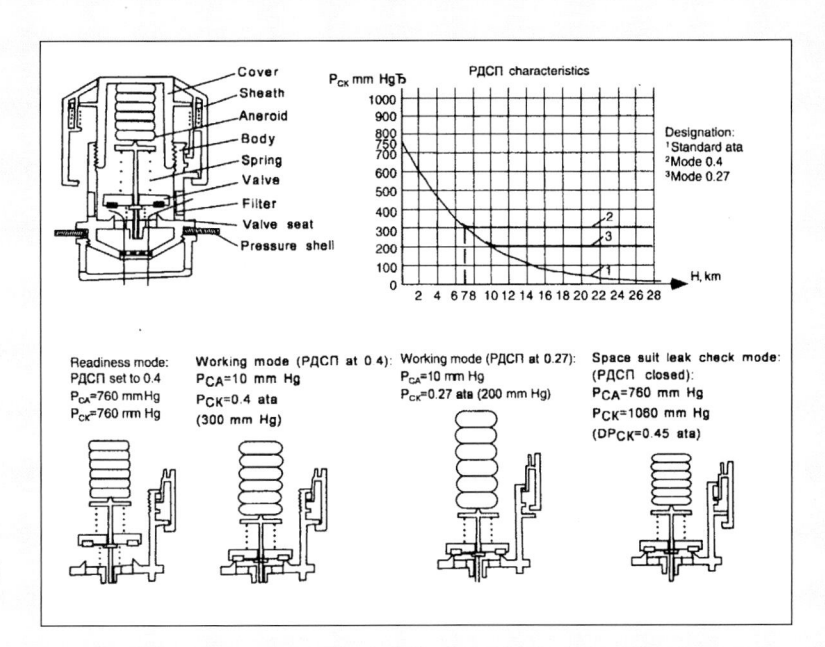

ABOVE The suit pressure regulator ensures that the suit itself remains at the appropriate pressure should pressure be lost in the descent module. The regulator is designed to maintain selected pressure levels lower than 6.4lb/in² (330mm Hg). The graph shows the relevant pressure levels with altitude and the respective settings on the controls. *(Energia)*

BELOW The sealed rubber suit gloves contain a fingerless pressure restraint shell with a locking connector to the arms of the suit. *(Energia)*

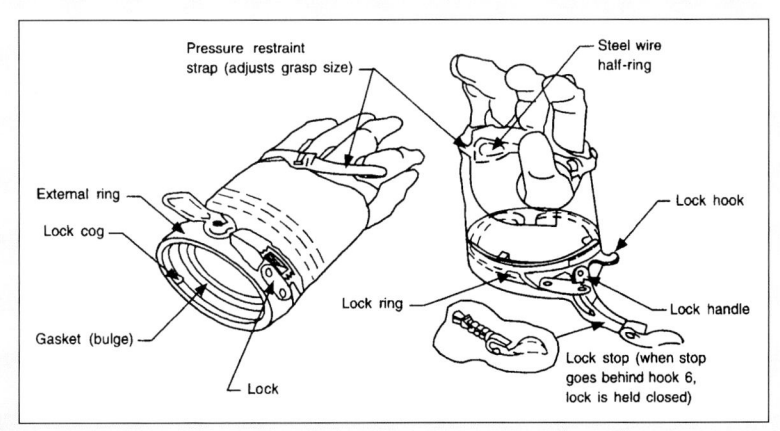

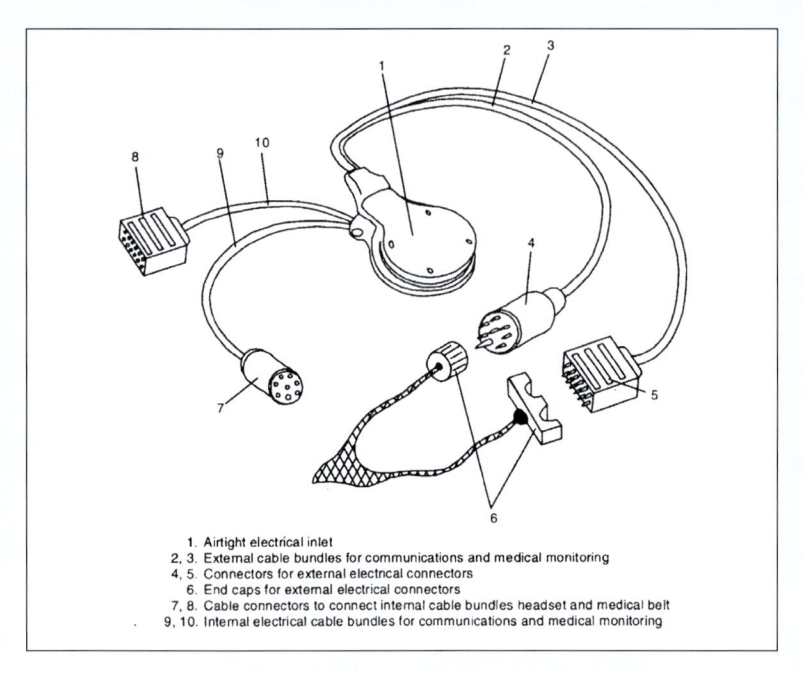

1. Airtight electrical inlet
2, 3. External cable bundles for communications and medical monitoring
4, 5. Connectors for external electrical connectors
6. End caps for external electrical connectors
7, 8. Cable connectors to connect internal cable bundles headset and medical belt
9, 10. Internal electrical cable bundles for communications and medical monitoring

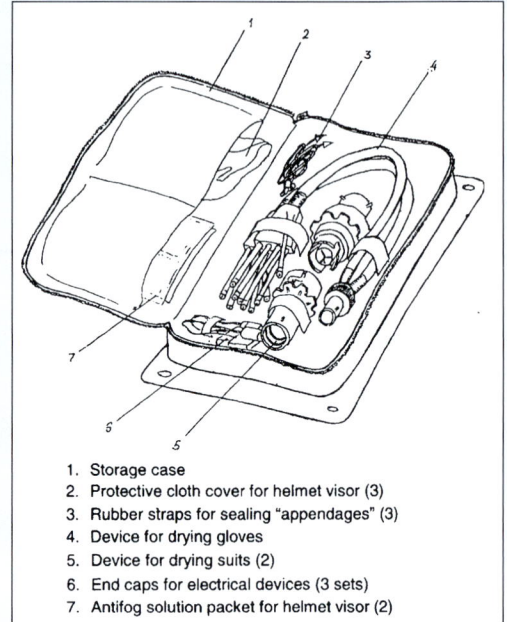

1. Storage case
2. Protective cloth cover for helmet visor (3)
3. Rubber straps for sealing "appendages" (3)
4. Device for drying gloves
5. Device for drying suits (2)
6. End caps for electrical devices (3 sets)
7. Antifog solution packet for helmet visor (2)

ABOVE Electrical equipment is required for the cosmonaut's personal communication kit and for the biomedical harness for monitoring of blood pressure and heart rate. *(Energia)*

ABOVE RIGHT The accessory kit contains two devices for drying suits when connected to the fan circuit in the spacecraft, a device for similarly drying gloves, three protective helmet visor covers, an anti-fogging solution for the suit helmet, three end caps for electrical connectors and three rubber straps for sealing loose cables. The kit is located in the orbital module attached to one of the panels near the locker. *(Energia)*

necessary to convert the suit pressure to pure oxygen to accommodate the material limits of a flexible suit while maintaining the essential levels of oxygen to the human body.

Propulsion

Soyuz TMA spacecraft have 26 thrusters, of which 14 each have a thrust of 29.3lb (130N) for approach and vehicular orientation. These are known as DPO-B thrusters and have a flow rate of 0.12lb/sec (0.05kg/sec). The remaining 12 thrusters each have a nominal thrust output of 6lb (26.7N) and are known as DO, for attitude control and orientation. They have a propellant flow rate of 0.02lb/sec (0.01kg/sec). Improvements to thruster performance and specific impulse have characterised the various evolutionary versions of Soyuz, and the first significant improvement came in 1998.

In the event that the retro-rocket fails to fire, four DPO-B thrusters can fire simultaneously to provide sufficient energy for the spacecraft to start down on its re-entry trajectory. On early Soyuz models the propellant was hydrogen peroxide, but Soyuz T and TM variants use nitrogen tetroxide (N2O4) as the oxidiser and unsymmetrical dimethyl hydrazine (UDMH) as the fuel. Collectively, these 26 thrusters are more commonly known today as the berthing/attitude control thrusters.

Later Soyuz T and TM/TMA versions integrated the propellant systems for the DPO-B and DO thrusters and for the Orbital manoeuvre engine (OME), which is used for orbital trajectory changes and for the de-orbit burn. This is known as the combined propulsion system, with the total propellant carried being available to all three rocket motor systems. Designated KTDU, the OME has a nominal thrust of 660lb (2.94kN) and can be fired 40 times in space. Total propellant flow rate is 2.36lb/sec (1.07kg/sec). The OME is mounted on a gimbal in the propulsion section of the instrument module and, with actuators powered by electric motors, can control thrust vectors in pitch and yaw up to +/- 5° of the spacecraft longitudinal axis. Total propellant capacity for the Soyuz TMA combined propulsion system is 970–1,967lb (440–892kg) depending on the specific mission.

The descent module has eight hydrogen peroxide rocket motors for attitude control during re-entry, each with a thrust of 22lb (98N). The system is pressurised and electrically enabled 14 seconds prior to separation from the truss structure separating the DM from the instrument module. Control authority switches to a landing computer program when the aft heat shield is jettisoned. The hydrogen peroxide propellant is brought to a catalyst under nitrogen pressurisation, where it decomposes to produce thrust. Control of the thrusters during descent is selectable between automated and manual control.

The recovery system consists of four parachutes (one of which is a backup), the soft landing thrusters, the reconfiguring of couch positions for impact and the automated landing system, which involves the deployment of communication equipment. During descent, the landing sequence is triggered by barometric pressure at an altitude of 16.9 miles (27.2km). Under normal operation, the pilot parachute deploys first with a dual canopy, automatically ejected when the cap to the pressurised container is jettisoned.

The bottom of the heat shield is jettisoned at an altitude of 8.8 miles (14.2km) exposing the base of the module and the six braking thrusters. The braking parachute is deployed at an altitude of 14 miles (22.5km) and a descent velocity of 515mph (828kph), slowing the DM to 200mph (320kph), at which point the 10,764ft² (1,000m²) main parachute is released to further slow the module to a descent speed of 14.5mph (23.4kph). During this period the seats are unlatched and raised to a suspension position that will afford some degree of shock absorption on impact with the ground.

To prepare the DM for landing, a gamma-ray altimeter issues a landing event signal at a height of 49ft (15m) from the surface, and at 2.6ft (0.8m) above the ground six solid propellant thrusters situated at the base of the module fire to brake the descent speed to a mere 4.5mph (7.2kph). This characteristic sign of Soyuz having reached Earth safely is evident in pictures of the landing where the sudden explosion of smoke from beneath the spacecraft creates a dramatic image. In the event of a malfunction to the primary recovery system, a 6,350ft² (590m²) backup parachute is deployed, which reduces the landing velocity to a descent speed of 22.4mph (36kph).

Flight data file

The onboard flight data file (FDF) has undergone some modification over the years. It provides each crew member with procedures and protocols for operating the spacecraft systems during normal or contingency operations. Up to the beginning of space station operations it carried over the formats from the Mir space station in which a 'Nominal modes' procedures book and an 'Off-nominal situations' book is available for each of the three crew members, while the 'Backup modes' reference book is available for the commander and the flight engineer. A single volume incorporating 'Reference materials' and a separate volume incorporating the 'Mission flight plan' is placed on board for availability to everyone. In all, ten volumes of five different books were installed within the descent module prior to launch.

In 1999 changes occurred which rearranged the types of book carried, now including separate volumes covering 'Launch/injection descent', 'Orbital flight', 'Off-nominal situations', 'Backup modes', 'Reference materials', and 'Flight plan', expanding to six the total number of separate volumes available in the spacecraft. The 'Launch/injection descent' book covered all activity in the spacecraft including pre-launch preparation, launch and orbital insertion, spacecraft depressurisation, transfer-hatch closure, pre-undocking operations, undocking activity, nominal and contingency de-orbit, emergency retrofire in the event of a main engine failure, and post-landing activity.

The 'Orbital flight' book now contained rendezvous procedures, automated approach to the orbital target, docking, pressurisation tests, countering a fire within the descent or orbital modules, transfer-hatch opening and general operations in the docked configuration. 'Off-nominal situations' covered depressurisation and fire within the habitable modules, dealing with failure in selected systems critical to function, and emergency procedures for more general failures. The other three books remained the same as with the earlier Soyuz missions.

Checklists are a crucial part of almost every

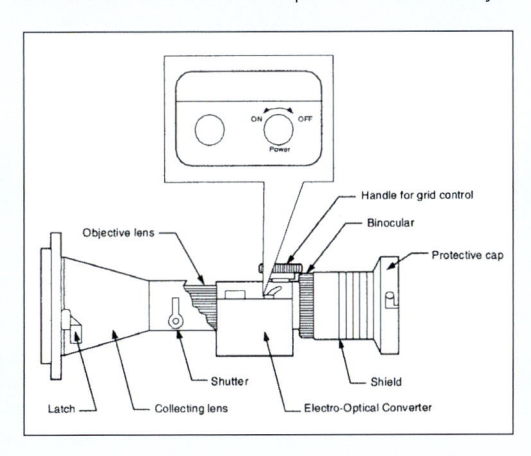

LEFT A night vision device is used for visual target acquisition and for attitude monitoring during approach and docking in the night portion of the orbit or during poor visual conditions. It has a focal length of 38mm with an aperture of 1:1.8 and a field of view of 20°. *(Energia)*

activity within Soyuz, no more so than with operational activities on the International Space Station and in NASA's Space Shuttle, when that was flying. Detailed step-by-step procedures assembled through extensive simulation on the ground responds to possible problems that may arise in orbit and usually requires the work of two crew members – one to read off the checklist and verify the actions completed and the other to perform the steps required for a given function. Very little is left to spontaneous decision on the spot and usually there is no physical function required.

A philosophical difference in Soyuz is that a gradually increasing degree of autonomy was handed to the crew, the realisation growing that not every part of the flight plan could be commanded from the ground or uplinked to the crew through verbal instructions. It was in Soyuz that the 'piloting' role of space flight emerged fully within the Russian system, which had been reluctant at first to surrender control to the cosmonauts on board. It was here that the shift from ground-based control to crew decision structures evolved, a critical part of maintaining a healthy spacecraft and, in its very essence, a justification for having people along.

Backup instructions are a means of providing the crew with a set of procedures for correcting processes that fail to go strictly according to expectations. Here, a switch configuration may not produce the expected reaction with spacecraft systems, or a set of actions on board may result in an unexpected result in the way the spacecraft responds. So a set of alternative procedures are provided for the crew to reconfigure existing and healthy systems, for workaround actions that will produce the same result. All such backup operations will have been fully checked out in simulators before the flight and as such are cleared as safe and potentially productive.

Off-nominal contingency situations have also been anticipated through emergency drill and simulation, insofar as the full spectrum of unexpected situations can ever be planned for. Here, crew safety, stemming a propagating effect, retaining spacecraft systems operability and restoring normal operations come before resuming flight plan activity. In the Soyuz book, 'Off-nominal' is defined as 'a spacecraft failure or a non-operative systems parameter deviating from its nominal status which could result in flight plan reconfiguration or selection of back-up modes'. This volume provides guidance regarding mitigating paths the crew can take for a very wide variety of failures, some potentially catastrophic if not corrected.

The 'Reference materials' book is the most useful volume alongside the flight plan, containing detailed information on the design, operating logic and maintenance order and rules for spacecraft systems, equipment and instruments. It describes the way the systems operate and allows the crew to estimate projected performance based on displays and indicators on the display console and operating panels. It also provides block diagrams and circuit diagrams related to the control panels, with the technical actions required by the crew to diagnose a failure or an unusual reading on the displays.

The book is written with full awareness of the crew's familiarity with all the systems and equipment on board and recognises the very high levels of training required for qualification to fly a Soyuz spacecraft. Much studying and simulator work will have gone into the preparation of the crew for flight, as each may be called upon to help stabilise a contingency or emergency. This familiarity with spacecraft systems is quite separate from the training

BELOW The laser rangefinder is used for measurements of objects at a distance of 475–19,700ft (145–6,000m). *(Energia)*

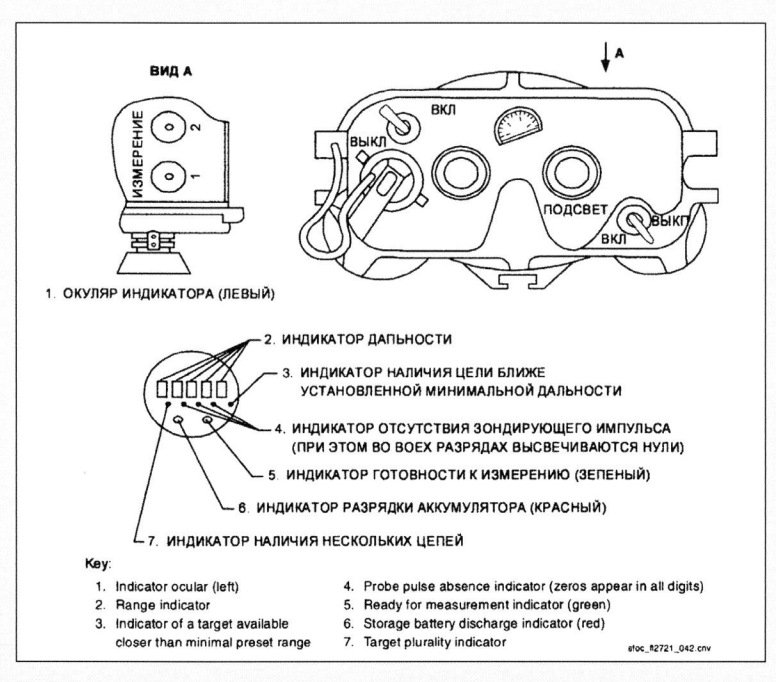

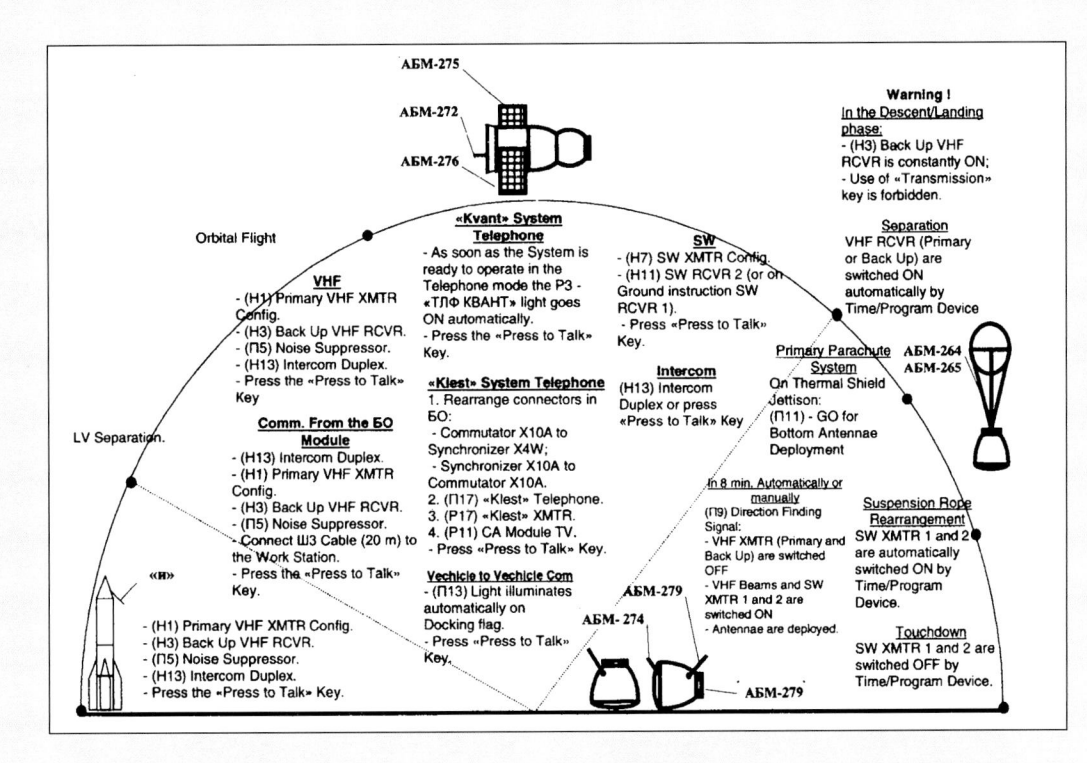

LEFT The various modes of communication between Soyuz and the ground tracking facilities are displayed on this chart for various phases of a typical flight including ascent, orbital flight and re-entry. *(Energia)*

required for a specific mission, the plans and procedures for which are identified in the 'Flight plan'.

The separate sections of the 'Reference materials' manual identify discrete sub-sets of systems or subsystems, on occasion where one subsystem may work in conjunction with another system. The first area identified is the motion control system, which describes the technical characteristics of the guidance and navigation systems. These involve attitude alignment, manual and programmed attitude control, description of manual controls and of manually calculating and inserting parameters for de-orbit operation, all of which involve propulsive burns. Also described are the rotation and translation hand controllers.

The next section describes the manual data entry unit, essentially a computer keyboard for inserting data into a programmable memory command system. It covers the full range of control words and display formats and guides the crew through sequences of commands and 'word' instructions to set up systems for independent operation or autonomous control. The actions resulting from this range from programmable rendezvous manoeuvres to the de-orbit manoeuvre at the end of a mission.

The section covering the attitude and orientation thrusters and the main engine, all situated on the instrument module in the combined propulsion system, provide reference information for the crew, while the descent reaction control system covers the thrusters activated for use while the descent module is re-entering the atmosphere. Similarly, the docking and internal transfer equipment is covered, with details of how to partially disassemble and reconnect elements of the docking and latching equipment that holds Soyuz fast to the docking interface at the International Space Station. All remaining sections covering spacecraft systems discussed earlier are contained in remaining sections of this volume.

Because the Soyuz spacecraft is an international crew transportation system, language has been a crucial element in selecting a common set of symbols and notation that can be readily understood by crew members from many countries. Russia has been flying foreign nationals since 2 March 1978 when it carried Vladimir Remek, from what was then Czechoslovakia, aboard Soyuz 28 to its Salyut 6 space station, where he remained for six days.

First in a series of Intercosmos missions, Russia would become familiar with the very different cultural backgrounds of its international crews, more than 20 years before assembly of the International Space Station began. In a very real sense, this reflects the multicultural nature of Russia itself, a collegiate of countries, nations and races that for centuries have learned each other's languages and different ways. For this reason, particular attention is paid to the written script and the notation applied to spacecraft systems, both aboard the vehicle and in the books carried in the flight data file.

The FDF is completed for a specified mission two months before the scheduled launch, about one month prior to completion of the crew training flight examination. Three weeks before launch, the Gagarin Centre outside Moscow receives two sets of the FDF, one set for the prime crew and one for the backup crew. The pre-launch period begins 10–14 days prior to launch so that the crew have the option to identify queries or make their own personal notes against specific operations in the timeline.

About two days prior to launch the refined trajectory information is entered into the primary crew's FDF and the writing materials cleared for use in space are fixed to the inside of the FDF. This is then handed over to the engineers who will install all but the 'Flight engineer's backup modes' book and the 'Flight plan' volume in the descent module, these latter volumes being placed in the payload containers in the orbital module. After ingress around 2.5 hours prior to lift-off, the crew extract the 'Nominal modes' book and continue with their integration procedures from that point using this material.

Pre-lift-off procedures call for visual inspection of the orbital and descent modules by the crew, to verify equipment locations pertinent to that specific mission. Seated in the couches, the hoses and connectors are hooked up to the suits and the ventilation system is switched on. After communication checks are secured with Launch Control and with Mission Control in Moscow, the lithium hydroxide cartridge is initiated for purifying the atmosphere of exhaled CO_2, and the closeout crew conduct final switch and lock positions before exiting the spacecraft through the orbital module, closing the hatch to the descent module.

With the ground crew exiting at T-2 hours, the crew set the onboard clock, monitor the propellant meter in the combined propulsion system and report indicator readouts and positions throughout the spacecraft. Once again, radio checks are conducted with Baikonur and with Moscow and the switches on navigation sensors are set to assigned positions. Then the seat belts are tightened and a pressurisation check carried out, ground control closely monitoring any variations – suspected faults at this stage will result in a postponement in the launch.

As the count continues toward launch the crew continue monitoring suit temperature and pressure and report the humidity and pressure levels in both habitable modules. Written records are made in the appropriate documents as a pseudo log of events and activities associated with the various spacecraft systems. Throughout Russian space flight history, compared to the US, less emphasis has been placed on electronic logging and much more reliance placed on completed checklists and written records.

With a pedigree going back to the early 1960s, the traditions that underpinned early practices with the Soyuz spacecraft have been retained to a much greater extent that would have been the case with a succession of different vehicle designs. In the US human space flight programme, each new vehicle (Mercury, Gemini, Apollo, Shuttle, Orion) has been a step up in capabilities throughout the design of systems and subsystems, with much less evolutionary migration between vehicle types than is the case with Soyuz, where the flight operations record goes back almost 50 years.

Flight crew

Over time, as the role of Soyuz has evolved, the categorisation of crew members too has changed to match the requirements of the period. Standardised now as a three-person spacecraft, Soyuz carries a commander and two flight engineer positions. During the Mir period, the second flight engineer position was designated cosmonaut-researcher, or simply passenger. The commander occupies the centre seat with access to all the control systems and with the most commanding view of all the instrument displays on the main console. The first flight engineer occupies the left seat, the second flight engineer the right seat.

Only Russian citizens and representatives

of the Russian Space Agency can qualify as a Soyuz commander. While command of the International Space Station switches from European to Russian and American astronauts and cosmonauts in sequence, command of the Soyuz spacecraft during all portions of independent flight is with the Russian spacecraft commander. He alone is responsible for the supervision of crew procedures, making decisions regarding unexpected situations and taking control of manual operations, and is the primary crew member for communication with Mission Control in Moscow.

The first flight engineer monitors systems performance during the flight, observes the activities of automated systems, prepares data for contingency operations in the event of a malfunction, and looks after the life support systems. This position can be filled by any astronaut or cosmonaut from any participating country. The second flight engineer is also responsible for some of the life support equipment and associated systems and carries out support functions for the other two crew members.

Launching Soyuz

Flight operations for reaching the Russian space stations in the Salyut/Mir series were worked out, with minor modifications, over several years of sustained operations. Although the procedures have varied over time, and until quite recently, the standard rendezvous template was adopted for Mir and for flights to the International Space Station.

The precise launch date and time were dictated by the position of the target vehicle in orbit and by lighting considerations on revolution 34. This was to ensure coincidence of the two orbit planes. In orbital dynamics, plane changes are very costly in the fuel they require to provide the necessary ΔV, so the alignment of the plane of the chase vehicle with that of the target is best done when the two coincide over the launch site. Moreover, any shift in launch date means that for every 24-hour delay the launch time will move back by 24 minutes so as to maintain the appropriate lighting conditions.

The launch phase to orbit lasts approximately 8min 46sec, with the first eight seconds being the period that the launch vehicle is held on the pad while computers analyse myriad parameters from sensors across the rocket. The rocket begins to pitch over on to its appropriate heading just 20 seconds into flight.

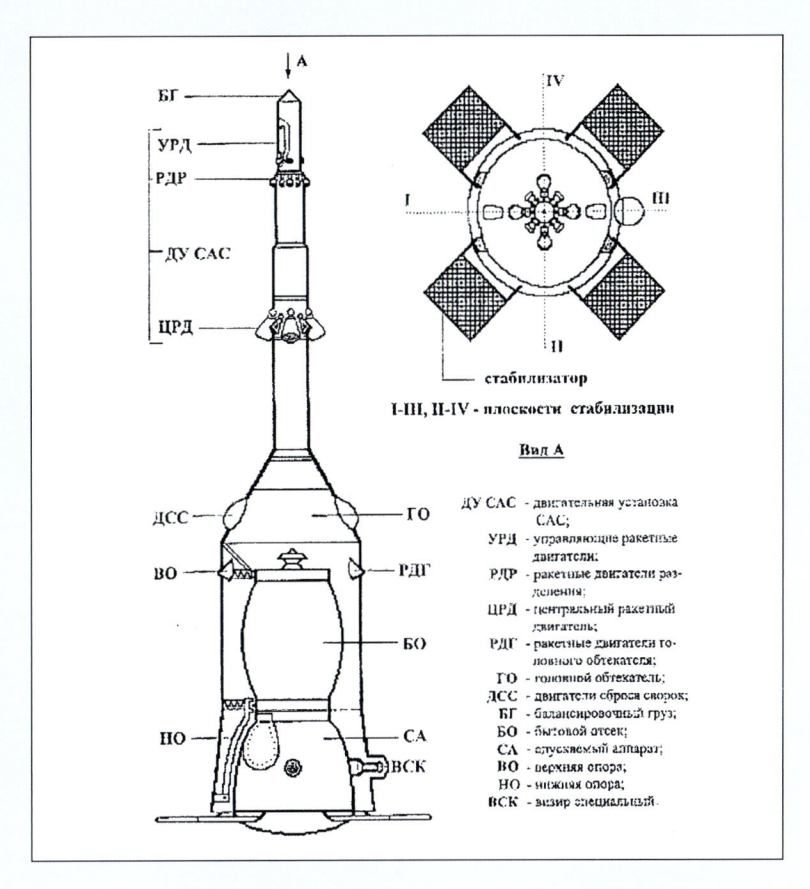

The maximum aerodynamic pressure on the launch vehicle must be kept at no greater than 757.8lb/ft² (36.28kPa), which affects the dynamics of the thrust phase and the propulsive burn of all rocket motors on both stages. This is known as max Q and is reached at 1min 5sec at a velocity of 1,017mph (1,638kph). At this point the vehicle is at a height of 6.9 miles (11.1km), 10 miles (16km) from the launch site.

The four boosters burn out at 1min 58sec into the flight, at a velocity of 3,490mph (5,615kph), an altitude of 25.8 miles (41.5km) and a range distance of 24 miles (39km), with acceleration at 3.5g. Maximum loads on the crew rarely exceed 3.5g, which is not a great deal higher than the loads which were experienced on a Shuttle flight. The launch escape propulsion system (LEPS) is fired away at 2min 1sec into the flight, followed 35sec later by the shroud encapsulating the spacecraft, which protects it from thermal and atmospheric

ABOVE The launch escape system has only once been required to lift spacecraft modules away from potential disaster, during preparations for the flight of Soyuz 10-1 on 26 September 1983. In the event of an abort on the launch pad, the escape motor would lift the descent and orbital modules to 3,290–4,920ft (1,000–1,500m), accelerating the escaping spacecraft to a velocity of 335mph (150m/sec). *(Energia)*

RIGHT Successive developments of the second stage for the Soyuz launch vehicle have provided increasingly greater capability, the RD-0110 being the rocket motor for the Soyuz-FG used to fly the TMA series. It is worth considering that this motor, with a thrust of 990kN, is more than 18 times as powerful as the RD-0109 that powered the Block E second stage of the Vostok launcher in the early 1960s. *(David Baker)*

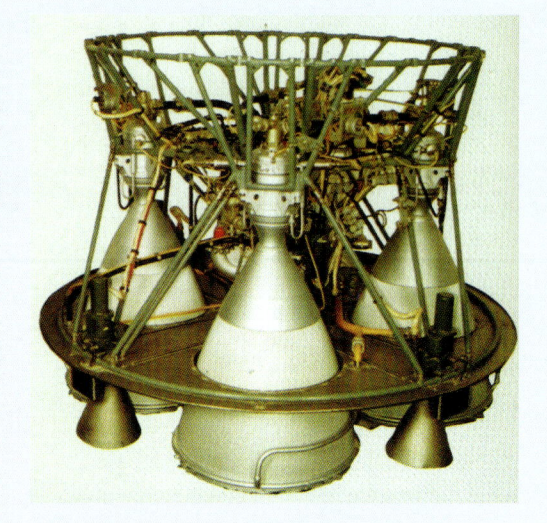

pressure loadings during flight through the dense lower layers of the atmosphere.

The upper stage ignites at 4min 45sec when the core stage is dumped, the vehicle now travelling at 8,232mph (13,245kph) and at a height of 104.4 miles (168km), with a downrange distance of 260 miles (418km). The acceleration loading is now a more comfortable 2.5g, but with the core stage gone the thrust/weight ratio is greater and the vehicle will begin to accelerate to a maximum 3.2g at second stage cut-off.

The second stage burns out at 8min 46sec at a velocity of 16,760mph (26,966kph), an altitude of 129 miles (208km) and a downrange distance of 994 miles (1,600km). Just three seconds later the Soyuz spacecraft separates from the third stage and is in independent flight. Nominal orbit injection parameters provide an inclination of 51.6°, an apogee of 149 miles (240km) and a perigee, or low point, of 125 miles (202km).

With these values, the spacecraft would remain in orbit for 301 revolutions before degrading into the atmosphere. But this is only a nominal value, the precise orbital lifetime being dependent on the 'stiffness' of the atmosphere, the precise pressure variations on the outer edges of the thin veil that encapsulates the Earth. In any event, further orbital manoeuvres required for the rendezvous procedure will modify these values and extend the 'life' of the orbit. All manoeuvres are calculated so as to provide sunlight for the final approach and docking to the International Space Station. For details of the history of rendezvous and docking in Russian spacecraft, see the separate feature 'Rendezvous and docking'.

Rendezvous operations

Orbital manoeuvres are set up when the correct phase angle is achieved. This is the angle subtended by a line from the spacecraft to the centre of the Earth and from the ISS to the centre of the Earth. These two lines will be widest at the point the Soyuz enters orbit and will continue to narrow as the spacecraft catches up with the station until they converge at the point of final rendezvous. During operations with the Mir space station, the phase angle could be anywhere between 240° and 30°, determined by the propellant capacity aboard the Soyuz spacecraft, which determines the magnitude of orbital manoeuvring to catch up with the target.

In the standard plan, rendezvous would be achieved on revolution 34, allowing plenty of time to plan the converging phase and altitude parameters necessary to bring the two vehicles together at the same time. On the second or third revolution the Soyuz would normally conduct a small manoeuvre to set up the phasing relationship to the target. A further manoeuvre takes places on revolution 17 or 18 to compensate for atmospheric drag. The Kurs-A rendezvous radar system would usually acquire the target at a range of 60 miles (97km), at which point updates are made to the relative parameters.

The three-phase rendezvous manoeuvre begins on revolution 32 with a transfer burn of 33mph (54kph) so that the Soyuz will achieve the same altitude and zero phase angle at the high point of its elliptical path. At a distance of 1,312ft (400m) the range rate is down to no more than 5.6mph (9kph), with the approaching Soyuz brought to a relative stop at a distance of 492ft (150m). At this distance the fly-around begins to align the spacecraft with the appropriate docking port, and the approach phase ends with berthing at a rate of 0.5mph (0.9kph).

Before flight, the trajectory will have been set up so that the docking target will be illuminated by the Sun from below at an angle of 30–60°. This is to ensure that should the Kurs-A automatic docking system fail the cosmonauts can see the target and manually guide the Soyuz to a hard dock. A wide range of options for setting up the automatic rendezvous and docking sequence is available to mission planners, with several stages before and after Kurs-A acquisition where the cosmonauts can intervene and update the guidance and navigation systems.

It is possible to set up rendezvous trajectories to approach the target from below and in front rather than from below and behind, or to conduct a range of different manual approach and docking modes. Some advantages are to be had from conducting a braking manoeuvre from a point ahead of the target rather than accelerating to conduct a docking from behind. The most complex manoeuvres are when an approach is conducted from some distance either side of the target, in which case the line of approach taken by the Soyuz spacecraft is at right angles to the station.

Returning to Earth

At the end of the mission, the descent trajectory is usually aligned with the first three revolutions of the recovery day in question, where the ascending node crosses the Equator at 20°E in longitude. The second parameter, the latitude, is open to planning options and is contingent on the ground track of the descent orbit. A range of possible landing sites is available within an area under what is referred to as the 'target line', and here the longitudinal positions lie between 63–74°E.

The lifting characteristic of the descent module is such that the spacecraft can 'fly' a distance of 47 miles (75km) left or right of the ground track. This allows for corrections during descent through the atmosphere for dispersions in the precise measurement of the orbit, or it can be used to move the spacecraft away from particularly poor weather conditions, which arise quickly and after the de-orbit preparations have commenced. With ground landing rather than splashdown, safety in these geographically challenging areas is paramount. Within a short distance, lakes, rivers, tall trees and small residential conurbations can pose serious and life-threatening hazards. The crew train extensively for post-landing activities, which include survival in remote and geographically hostile regions and in remote places where recovery forces may be several days in reaching the crew. Any malfunction in orbit can carry the spacecraft very great distances from the planned recovery zone and there have been many examples where crew members have had to survive Arctic conditions in winter until the arrival of recovery forces.

Standard recovery procedures for separating from the Mir station begin with the crew returning to the descent module and sealing the hatches – one at the docking interface and one separating the orbital module from the descent module. Thrusters complement springs, which impart a separation velocity of about 0.3mph (0.5kph), at a distance of 80ft (25m) from the station. An 8sec burst imparts a velocity of 1.1mph (1.8kph), which places the Soyuz above and behind the station after 1.5 revolutions of the Earth (about 2hr 15min).

For returning from the International Space Station, the Soyuz executes a slightly different set of manoeuvres that culminate – as they did with Mir – in a de-orbit burn to begin the descent towards the atmosphere. The magnitude of the burn depends on the altitude of the station. At a height of 124–186 miles (200–300km), the Soyuz main propulsion system will produce a velocity change of 200mph (322kph); from a height of 186–205 miles (300–330km), the velocity change will be 229mph (368kph). At this maximum altitude or slightly above, the velocity change will be a braking burn reducing the speed of the spacecraft by 258mph (415kph).

BELOW No other space vehicle type has endured as long as Soyuz. TMA-19 is docked to the Rassvet module on the International Space Station, reflecting the important role it now plays as the sole means of getting people to and from this permanently manned facility. *(NASA)*

95

THE SOYUZ SPACECRAFT

ABOVE NASA astronaut Nicole Stott sizes up the Soyuz descent module, epitomising the truly international crews that now fly in Russia's flagship spacecraft. *(NASA)*

BELOW The Soyuz TMA-19 spacecraft with NASA astronaut Doug Wheelock (left), Expedition 25 commander, and fellow crew members NASA astronaut Shannon Walker (right) and Russian cosmonaut Fyodor Yurchikhin, demonstrating the snug fit inside the descent module. The three crew members were returning from over five months on the International Space Station, where they served as members of the Expedition 24 and 25 crews in 2010. *(NASA)*

In the early phase of Mir operations and with the Soyuz TM spacecraft, the orbital module was jettisoned, but around the time activities transitioned from the Mir station to the International Space Station in the late 1990s the OM was retained until completion of retro-fire. This resulted in a deceleration burn of 4min 20sec to reduce speed by 258mph at the altitude of the ISS.

Approximately 22.5 minutes after the de-orbit burn, when the spacecraft has descended to an altitude of about 87 miles and is very close to entry interface with the atmosphere, the orbital module and the instrument module are separated by pyrotechnic devices. The instrument module moves away tangentially to the re-entry path and separates at around 1.3mph (2.1kph), the orbital module departing with a speed of 1.8mph (2.9kph). The descent module will reach the ground about 17 minutes later.

Immediately after the separation of the redundant orbital and instrument modules, the descent module attitude control logic holds the spacecraft in roll and yaw with pitch commands limited to 2°/sec. The attitude control system will have been set up prior to re-entry such that roll and yaw channels are held on a free 3°-of-freedom, orientated such that the axis of rotation is perpendicular to the orbit plane. The gyroscopes are uncaged at 1min 14sec before separation and the attitude rate sensors are powered to measure and control the pitch rate of the spacecraft.

After separation, the crew can visually monitor the slow rotation in pitch, and the spacecraft will usually make one or two full turns. This has the effect of turning the descent module around so that the base of the heat shield is facing forward. This orientation is established about one minute before entry at an altitude of 59 miles (95km), so that by the time the spacecraft encounters the atmosphere at a height of 50 miles (80km) it will be stabilised in a base-first attitude. At entry, the module is about 7.5min to 8.3min from landing.

For most of the descent phase the decelerations are around 4–5g, depending on the specifics of the mission and the flight parameters. This may not seem excessive, but for crew members returning from the International Space Station after a six-month expedition in weightlessness it's a lot! At peak deceleration the loads reach a maximum 8–9g, which produces noticeable effects on the crew members suddenly exposed to such high forces.

There have been a number of dramatic incidents involving Soyuz spacecraft returning to Earth following problems in orbit, but none have been fatal since the loss of Soyuz 11 in 1971 – a safety record of more than 40 years without loss of life. Nevertheless, some flights have ended in unusual circumstances, and on one instance the crew were left for several days without recovery in hostile conditions and adverse weather after enduring a re-entry that exposed them to almost 20g.

Despite the extended heritage of Soyuz hardware, and the age of the basic design, the spacecraft is one of the most reliable transportation systems devised by humans. It is flexible, adaptable and has sustained several generations of development into a wide variety of versions and variants.

ABOVE Soyuz TMA-7 floats free, displaying key features of the exterior, a product of an evolving design and engineering process over more than 50 years in which more than 120 spacecraft of this type have been launched into space since 1967. *(NASA)*

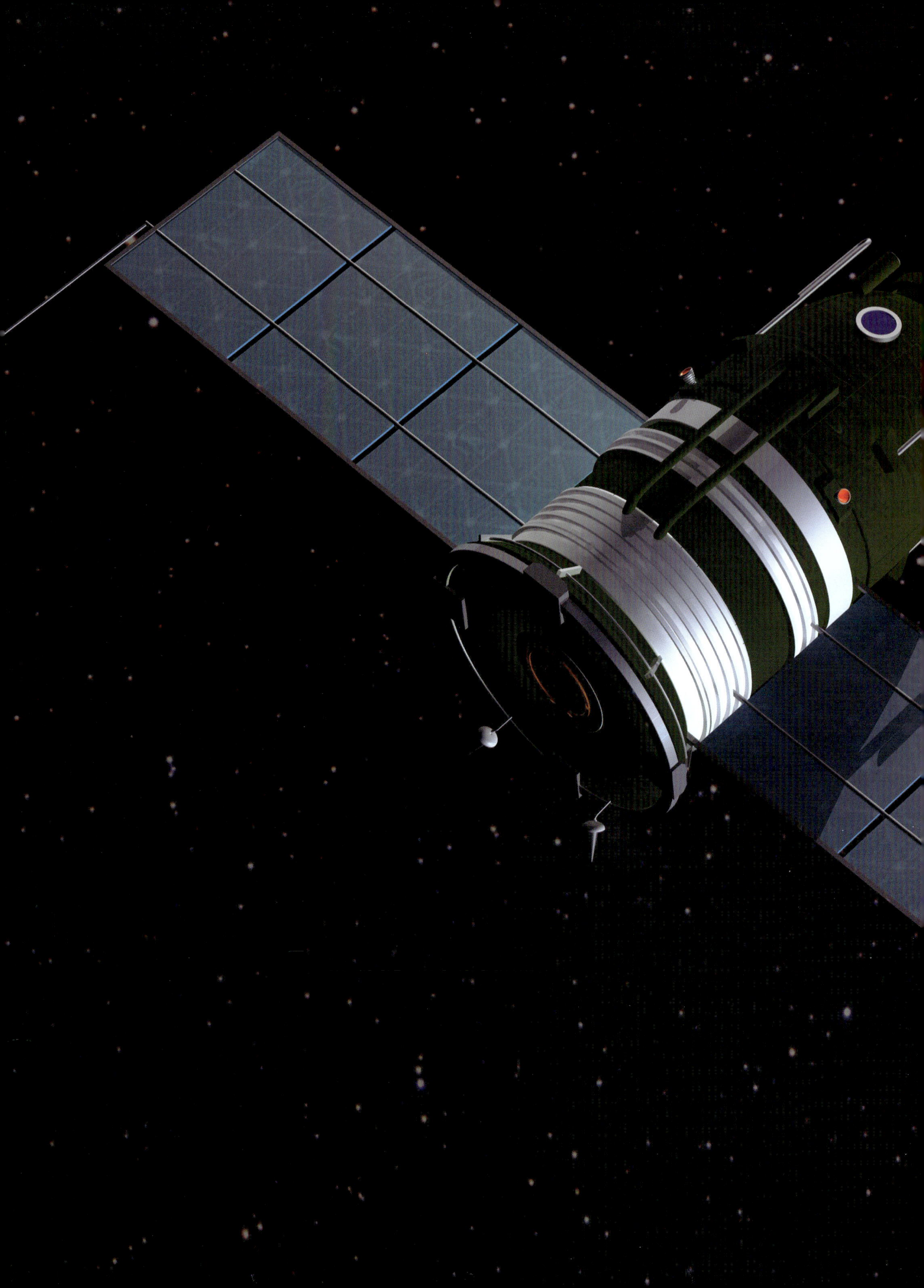

Chapter Four

Soyuz 7K-OK and Zond flights 1966–71

The first flight of a Soyuz spacecraft began with the launch of 7K-OK-2, designated Cosmos 133, by an 11A511 rocket on 28 November 1966.

OPPOSITE Zond pioneered several technologies which would be fed into the basic Soyuz programme, but these circumlunar flights were, in themselves, a tremendous tribute to the design of a multipurpose spacecraft with several potential applications. *(Nick Stevens)*

The plan was for the second spacecraft to be launched one day later and to achieve an automated rendezvous and docking for a period of three days during which all the spacecraft systems would be thoroughly checked out via the telemetry systems on board. Then the two spacecraft would separate and return to Earth a day apart. But no sooner had 7K-OK-2 reached orbit than it began to fail, pressure in the attitude and orientation propellant tanks and consumption being so high in trying to stabilise the spacecraft that it ran out of propellant within minutes.

There was now no purpose in launching a second spacecraft and that was cancelled. With further problems piling up the spacecraft was commanded to re-enter, but a series of failures to the propulsion system caused it to re-enter at too shallow an angle, which triggered a self-destruct system detonating 51lb (23kg) of explosive, destroying the spacecraft completely and sending its remaining fragments into the Pacific Ocean.

A review of the mission concluded that it would be prudent to launch the next spacecraft on a solo flight, but when that took place on 14 December the rocket erupted in flames, triggering the escape system, killing one person and sending the spacecraft (7K-OK-1) tumbling to Earth some distance from the pad. But if there were problems with Soyuz, the Americans were about to experience one of the most tragic events in the history of the space programme when, on the evening of 27 January 1967, NASA astronauts Virgil 'Gus' Grissom, Ed White and Roger Chaffee were killed after fire swept through their Apollo spacecraft.

There was a feeling of shared sorrow, the common humanity of good people sweeping aside for a moment the jaundiced politics of ideology, as scientists, engineers and technicians in Russia felt for the loss of these three American astronauts. In the days and weeks after that tragic event, there was a certain feeling of reprieve from the intense pressure the Russian teams had been under to beat NASA to a flight around the Moon. Little did they know that within a matter of weeks they too would be the recipients of condolences sent from the United States.

The next attempt to get a Soyuz in space came on 7 February 1967, when 7K-OK-3 was carried to orbit with the mission, designated Cosmos 140, destined for a two-day flight. Trouble again plagued the operation, the solar arrays stubbornly refusing to align themselves with the Sun, there were further problems with the propellant systems and a misdirected retro-burn put the spacecraft down on an ice floe in the Aral Sea.

On rushing to rescue it, recovery teams saw the descent module fall through the ice and sink! This was of serious concern to the design team because the DM had been built to remain afloat in water, and had been tested positively for such an emergency. Post-flight analysis showed that all the systems that had failed in orbit would not have threatened the safety of a crew and could have been circumvented with manual controls which would have been available on a manned mission. Moreover, all life-support equipment on board had worked well. There was a degree of confidence that with a little more testing and tweaking, the design was sound in principle and could be made to work.

The first launch of a lunar version of Soyuz occurred on 10 March 1967 when 7K-L1-2P was sent into space on the first of Chelomei's UR-500K Proton rockets, a graceful, sleek and high-performance launcher that worked to perfection. During the course of the flight the upper stage of the Proton fired, pushing the Soyuz spacecraft (known to the wider

ABOVE By the time the Voskhod flights had ended the Soyuz spacecraft had gone through a major change in layout, although the interior arrangement of three seats for unsuited cosmonauts experienced several variations. *(TASS)*

world only as Cosmos 146), travelled on an elliptical path out to a distance equivalent to the Moon's orbit and returned to burn up in the atmosphere. An attempted repeat on 8 April sent 7K-L1-3P (Cosmos 154) into low Earth orbit but no further into its planned mission due to a fault in the upper stage, which should have fired it out into deep space. Instead, Cosmos 146 decayed naturally out of orbit and burned up in the atmosphere.

The 7K-L1 lunar version was not at all representative of the main Soyuz variant being planned for Earth-orbiting rendezvous and docking and for crew habitation supporting the N1-L3 Moon landing operation and its LOK one-man lander. But the primary purpose of Soyuz was again hijacked by the political elite as the new leadership in the Kremlin sought to use the space programme for propaganda purposes. Concerns expressed by engineers over the readiness of Soyuz to begin carrying people were overruled by the direct demand from Brezhnev that May Day 1967 should be marked by something special from space.

Tragedy strikes

Notwithstanding the lack of a totally successful unmanned flight to qualify all the systems and the only partial success of Cosmos 140, it was decided to send the next mission up with cosmonaut Vladimir Komarov on board. There was direct pressure for a joint flight of two manned Soyuz launched separately, one carrying a single occupant, the other with three cosmonauts on board. Komarov would be the first Soviet cosmonaut to make a second flight.

The crew for the second flight would consist of Bykovsky, Yeliseyev and Khrunov. The two spacecraft would dock together on the first orbit, with two cosmonauts doing a spacewalk to move from one orbital module to the other, joining Komarov in his spacecraft with TV cameras showing the historic event. The mission was designed to achieve the first docking between two manned vehicles, the first dual spacewalk and the first transfer of cosmonauts launched in one spacecraft and returning in another. With all Apollo flights on hold as a result of the tragic fire in January, the Russian star appeared to be in the ascendant.

The flight of Soyuz 1 began when 7K-OK-4 was launched by an 11A511 from Baikonur on 23 April 1967. Reaching orbit, Soyuz 1 immediately ran into problems when one of the two solar cell wings refused to deploy and the 45K solar-stellar attitude orientation system malfunctioned. With this out, telemetry antennae not deployed and his spacecraft unable to point the remaining solar panels at the Sun, a rendezvous and docking was out and the second flight was stood down. But that was only the beginning of a dismal mission in which Komarov had to turn off as many systems as he could to conserve power. With a meagre 14A available from the batteries, his spacecraft would run out of power after a day.

Unable to satisfactorily orientate his spacecraft manually, there were serious fears that he would not be able to align the thrust vector of the retro-engine, and being out of radio contact with the ground between

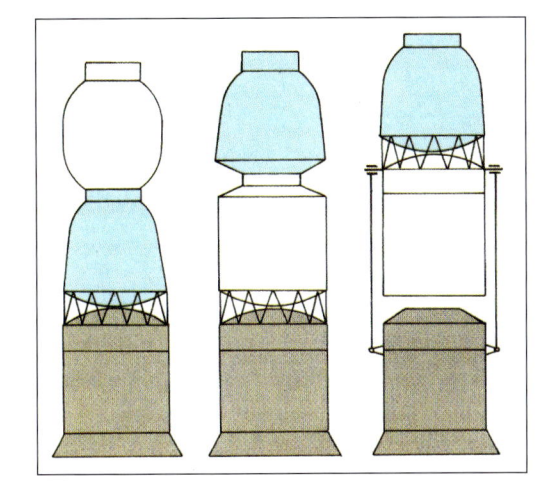

LEFT The descent module (blue) evolved to its final location with the general configuration of modules (left) from derivations which tracked the changing mission role envisaged for Soyuz. *(David Baker)*

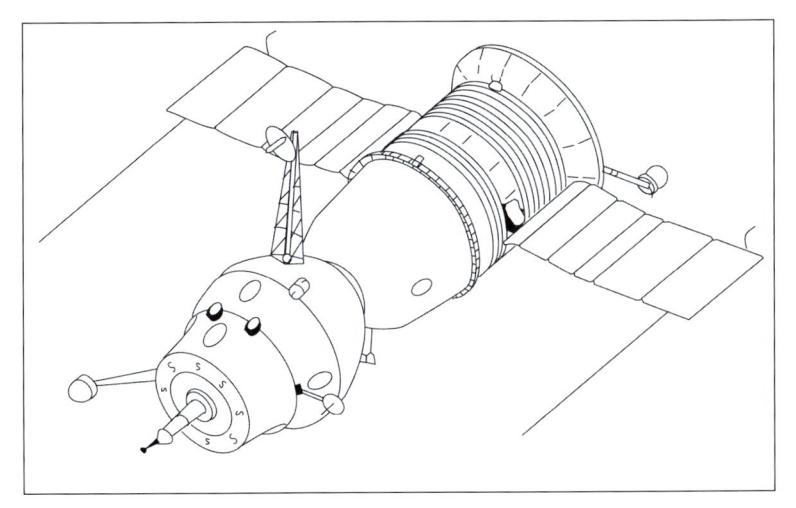

ABOVE The 7K-OK Soyuz became the baseline and from this would evolve the Soyuz of the 21st century. *(David Baker)*

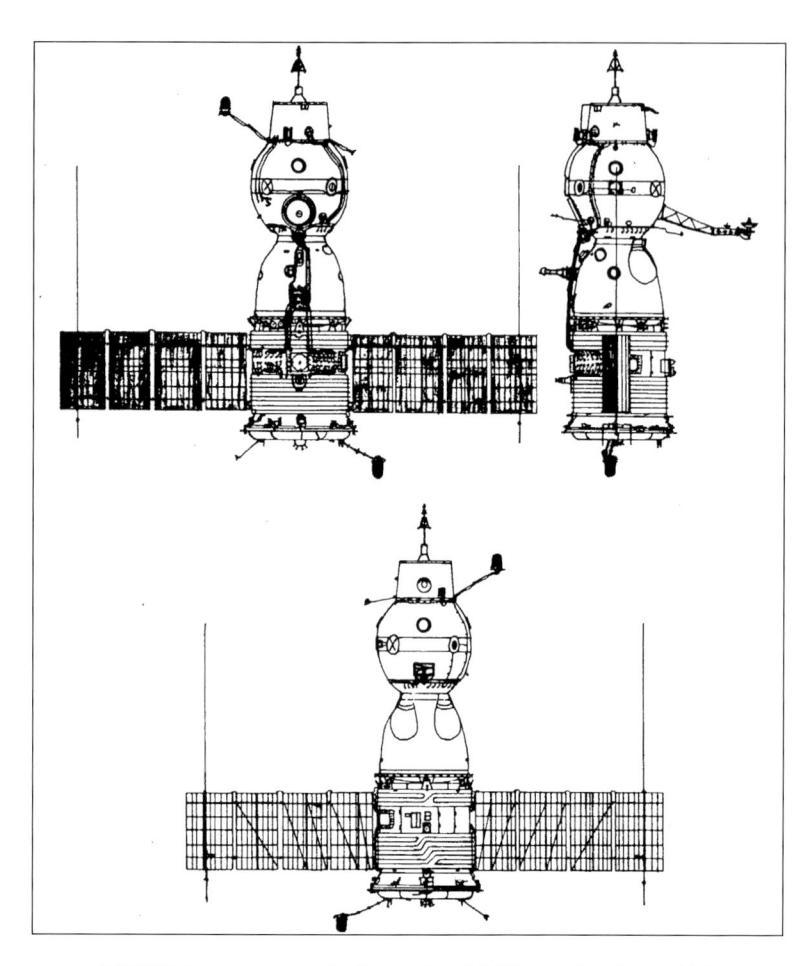

ABOVE Soyuz 4 represents the definitive first-generation flight vehicle that would be used with a variety of mission requirements, primarily for Earth orbit rendezvous and docking, seen initially as a means by which cosmonauts could be sent around the Moon. *(Charles P. Vick)*

ABOVE A production line of descent module structures comes together at the assembly plant within Mishin's facility. *(Charles P. Vick)*

orbits 7 and 14 did little to allay fears. When communications were restored as the Earth once again rotated to place the ground track of the flight path across the USSR, Komarov reported that even his ionic sensors had packed up. With frantic attempts on the ground to find a safe way for him to orientate his spacecraft, and with a flight control team that had not slept for 24 hours, Komarov was running out of time.

On the next two orbits Gagarin calmly read out the precise instructions for what was a highly complex manual orientation procedure using the KI-38 gyroscope and an eyeball alignment on the Earth's horizon at selected times – a procedure which had never been simulated on the ground, using techniques devised as the flight progressed. Komarov performed these impeccably and tried to reassure controllers on the ground that all would be well.

The retro-burn went ahead as well as possible under the difficult circumstances and communications broke off as the spacecraft began entry into the atmosphere and created the expected plasma sheath. When the time came to deploy the main parachute it stuck fast in its container, a malfunction that should have triggered its release and deployment of the backup. But because the backup was locked below the prime parachute inside the top of the descent module it could not be set free. Soyuz 1 fell to Earth and hit the ground at a speed of 90mph (144kph). Collapsing in upon itself, the impact ignited the 66lb (30kg) of hydrogen peroxide which should have been used in the braking rockets to slow the descent a second before landing.

In contrast to the expectations of its designers, control teams and the political overseers, this flight, which lasted a mere 26hr 48min instead of the planned four days, brought several days of national mourning, a State funeral and incarceration in the Kremlin wall and a sense of universal grief. This time, condolences were received by Russians who only three months earlier had shared commiserations with the families of three Apollo astronauts. A most grim reflection on a stunning

FAR LEFT Each descent module was virtually fabricated by hand, and several layers of quality control were involved in preparing a wide range of hardware for numerous tests and qualification flights, as well as spacecraft rated for flight. *(Novosti)*

LEFT The definitive Soyuz orbital module has lost its drum-like shape and assumed the appearance of an ellipsoidal barrel. *(David Baker)*

BELOW A training simulator at Star City where cosmonauts will learn how to familiarise themselves with the controls, displays and general layout of the orbital module, and where crews will practise for particular missions. *(TASS)*

tragedy was that had the earlier failures not scrubbed the second Soyuz with three cosmonauts, that would have gone into space with the same flawed parachute exit orifice, and four people would have died, not one.

In eerie parallel to Apollo, a manned Soyuz would not rise to fly again until October 1968. For 18 months three major Russian manned space flight programmes were stalled, including the Earth-orbit Soyuz programme represented by Soyuz 1, the L1 Soyuz (Zond) for circumlunar flights, and the L3 Soyuz for playing a role of mother ship to the manned Moon landing launched as part of a colossal stack sent into space by the N1. But there was another, very secret part of the Soyuz programme which remained hidden within the dark recesses of the Soviet military research and development effort.

The Soyuz gunship

While preparations for Soyuz 1 were still gearing up, on 24 August 1965 the Central Committee and the Council of Ministers signed a decree authorising production of a purely military Soyuz. Designated 7K-VI, the military Soyuz would reverse the arrangement of the orbital and descent modules, requiring a hatch to be installed in the base heat shield of the latter to permit entry into the OM. This had been seen as highly dangerous by the engineers who had nursed the basic Soyuz design through more than five years of evolutionary development.

But there was a precedent for this of which the Russians were only too well aware. Published details of an American military space station announced in 1963 included precisely this arrangement. By attaching a military version of NASA's Gemini spacecraft to the forward end of a cylindrical laboratory called the Manned Orbiting Laboratory (MOL) and launched on top of a Titan III rocket, the two-man crew would open a hatch in the base heat shield and

RIGHT The 7K-V1 Soyuz gunship with 23mm cannon in the front of the orbital module which was proposed as a means of defending manned space stations and other facilities in orbit. Not as far-fetched as it sounds – in America engineering analysis was made of NASA's Lunar Module to determine whether it could be used for offensive operations in space. Neither Russia nor America ever deployed such vehicles operationally. *(David Baker)*

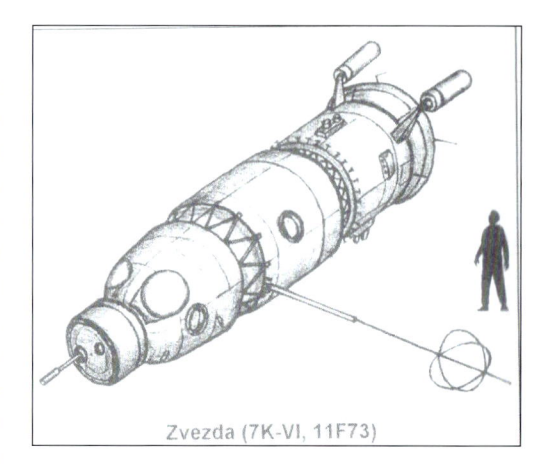

Zvezda (7K-VI, 11F73)

work inside the station. Responsible for the 7K-VI, Dmitry Kozlov made significant changes to the design of the basic Soyuz, providing a wide range of equipment in the orbital module which would remain attached to the equipment section when the descent module separated after retrofire. The military Soyuz would be called Zvezda, not to be confused with the module of the same name used as the fundamental building block of the Russian section of the International Space Station.

The design and layout of the Zvezda descent module would be modified to accommodate these changes, the two cosmonauts' seats being arranged in a splayed 'V' shape to make space for the centre floor hatch through which the crew would access the orbital module below. It would also be equipped with the Svinets device originally designed for, but never used with, the Vostok programme for monitoring the launch of ballistic missiles. It would also be equipped with a long extendible boom for sensors capable of collecting data on other satellites, on satellites approaching the Zvezda and for gathering electronic intelligence information.

With a weight in excess of 14,300lb (6,500kg), the launch vehicle would be upgraded further, and for this role the 11A511M was designed, although it was still recognisably similar to the R-7 which had already done so much for the Soviet space programme. While the publicly acclaimed Soyuz missions would proceed, garnering laurels for Soviet achievements, the military Soyuz would carry a variety of equipment for a wide range of flight roles, and with the ability to remain in space for up to 30 days this was a considerable expansion of the spacecraft's capabilities, bringing new possibilities.

In designing the electrical power supply for Zvezda, Kozlov dispensed with the potentially troublesome solar arrays, opting instead to equip the ship with radioisotope thermoelectric generators (RTGs), which in producing heat through radiated energy from the natural decay of plutonium would produce electricity. The plutonium fuel sources would be locked inside casks surrounded by thermocouples for the production of electrical power. This concept was not new and was being developed in America for providing electrical power to the scientific instruments Apollo astronauts would leave on the Moon. Like an orbiting spacecraft, half a 'day' on the Moon would be in darkness, when solar arrays are useless.

Not least of the military Soyuz's potential advantages was the photo-reconnaissance capability of this bigger and larger spacecraft compared to the Zenit, as was its orbit manoeuvring capability to carry out satellite inspection. Others were its, albeit limited, velocity capacity to significantly change the orbital period. This would allow for realigning the ground track for extended observation of sites of interest on the ground. But perhaps the most dramatic role for the 7K-VI Zvezda was that of an armed space fighter.

Under the engineering guidance of Aleksandr

RIGHT Covered in thermal protective insulation, and with solar panels folded for encapsulation in the launch shroud, a Soyuz is packaged ready for flight. *(TASS)*

RIGHT In orbit the solar cell arrays are deployed to their full width of approximately 33ft (10m), with the first-generation Soyuz having a gull-wing appearance that distinguished this version. *(TASS)*

CENTRE The launch escape system on early Soyuz 7K-OK spacecraft is similarly distinguishable by its snub nose and shorter tower than would be designed for later generations of Soyuz. *(Novosti)*

Nudelman, chief of the Design Bureau of Precision Machine Building, a space gun had been developed for use as an anti-satellite weapon or, in the case of pursuit of incoming non-Russian spacecraft, to destroy a spacecraft engaging in a close fly-by for 'snooping' on Soviet activities. It was also said to be useful for preventing the over-flight of Russian territory by US spy satellites, although given the physics of orbital dynamics quite how that could have been achieved in reality is difficult to understand.

The gun was built in Department No 3 at Nudelman's facility and tested on an air-bearing rig to see if it would have an adverse effect on the stability of a spacecraft in space. Cynics had suggested that in firing, the gun would impart a recoil action to the body of the ship, which would react by spinning round. This was found not be the case and, the engineering details having been solved, the physics was found to be supportive. An optical system installed in the descent module would allow a crew member to align the gun with the target and fire at will. Unknown until recently, even some of the 'civilian' Soyuz spacecraft carried elements of the firing sight designed for the gun as an engineering test of friction bearings in weightlessness, but no gun was ever taken into space on a non-military spacecraft.

In international frames of reference, a precedent for putting hostile weapons in space had been set when in 1960 the Americans began a programme called SAINT (Satellite Inspection). Before that, early in the 1950s, US astronaut Edwin 'Buzz' Aldrin, the man who would accompany Neil Armstrong on the first Moon landing, wrote his doctorial dissertation on the principles of orbital rendezvous and satellite interception. The first project studies

BELOW Unlike NASA's Saturn launch vehicles, which were stacked vertically in the launch position, the 8K72 series of Russian rockets were fitted together in the horizontal position. *(RKK Energia)*

RIGHT The assembled rocket is rolled on a rail line to the pad, where it is erected vertically into the cantilevered launch position. *(Novosti)*

FAR RIGHT In the vertical position, held by gravity within the four support arms, one of which can be seen with technicians in attendance, the vehicle is embraced by a pair of scissor arms containing multiple access platforms. These will be removed before launch, exposing the vehicle for flight. *(Novosti)*

began in 1957 with a determination to produce an anti-satellite system. This was tested on 13 October 1959 in a programme called Bold Orion, in which a missile fired from a Boeing B-47 bomber intercepted and passed close by the US science satellite Explorer VI. The remainder of this programme is still classified.

In June 1962 the US Air Force proposed a manned satellite interceptor and quickly focused on NASA's Gemini spacecraft for which the 'Blue' Gemini programme was formed. This was different to the Gemini B that from 1963 was adopted as the planned spacecraft to support the Manned Orbiting Laboratory. In other studies, the Air Force pressed for the use of the newly discovered laser beam as a weapon with which to threaten and destroy enemy satellites as well as incoming missile warheads.

In 1962 also, the US Navy demonstrated an anti-satellite capability with its Hi-Hoe programme and the US Army deployed the Nike-Zeus as an anti-satellite weapon, demonstrating its capability in 1963. That same year, the Air Force began development of an adapted Thor missile as an anti-satellite weapon under Program 437, a system capable of knocking out satellites at a height of up to 400 miles (644km). The first demonstration satellite intercept took place in May 1964. From this activity came a focusing of effort in satellite interception from ground-based missiles to a space-based capability.

The satellite interception function of the Air Force MOL programme was not, and has never been, admitted publicly, but an internal contemporary report concluded that the MOL 'is expected to provide the major realistic assessment of whether or not orbital inspection and interception schemes (with manned Gemini vehicles) are indeed feasible.' By 1965, Arthur Kantrowitz of the Avco Corporation was advocating the orbiting of laser and particle-beam weapons for destroying enemy satellites and ballistic missiles in flight. And as if to quantify the magnitude of the test phase, between 1963 and 1967 eight separate interceptions were made against phantom targets.

It was against this background of impressive and alarming developments from the US –

FAR LEFT Enveloped by its servicing and access gantry, the Soyuz rocket spends little time at the pad prior to lift-off. Launch preparations, and the flight itself, can take place in sub-zero temperatures and snowfalls. *(RKK Energia)*

LEFT The gantries are removed by pivoting away at 180° in opposition to each other prior to the flight. *(NASA)*

BELOW Most flights begin on the Start-1 pad, where Yuri Gagarin left for the first human flight into space. Cantilevered over the blast trench, flames will be directed out to the right. *(RKK Energia)*

about which a considerable amount must have been known to the Soviets through several 'open' sources in the West, not least the specialist press – that plans to siphon off some Soyuz spacecraft from the production line were made. Zvezda would be the Soviet equivalent of the Blue Gemini for military applications, some of which would have had hostile roles.

But Zvezda was not the only response to increasingly sophisticated developments in the militarisation of space from the US. Outside the scope of this book, but associated with the Soyuz spacecraft, in October 1964 Chelomei had proposed a large orbital complex called Almaz, studies indicating it to be a large space station called the Orbital Piloted Station (OPS) supplied by a version of Korolev's 7K-OK and designated 7K-TK.

Mindful that the Ministry of Defence was paying for Soyuz, in 1962 Korolev had proposed two military variants: the Soyuz-P, a piloted anti-satellite interceptor, and the Soyuz-R, a strategic reconnaissance version. Soyuz-R was in fact a small space station that would have been supplied and manned by ferry

OPPOSITE This engineering layout of the Soyuz launch vehicle represents the current generation of rocket and spacecraft. *(via Jim Gerrity)*

RIGHT Lift-off, as the thrust builds and the rocket lifts itself from the support cradle! This Soyuz is one of a later generation, as indicated by the extended launch escape tower. *(RKK Energia)*

BELOW The general layout of the pad and the disposition of the support arms, the servicing gantries and the umbilical towers are unchanged since the flight of Sputnik 1 in 1957. *(RKK Energia)*

flights with an adapted Soyuz spacecraft called the 7K-TK. To some degree the Zvezda version of 1966/67 was the successor to the Soyuz-P, while the Soyuz-R complex lent Chelomei the use of an existing manned spacecraft for his Almaz programme.

In 1966 Kozlov began work on the ferry-ship to Almaz, Chelomei's working relationship with Mishin being more harmonious than it had been with Korolev. But development work on the Almaz station was long and protracted, second to the circumlunar Zond programme and the N1-L3 Moon landing. Concerns over the perceived, but totally unrealistic, threat to their activities from military space technology in the US drove designers to adopt Nudelman's 23mm cannon for Almaz, installed as a defensive weapon and ineffective for offence due to the limited manoeuvring capabilities for the station. But it would be many years before any of this hardware made it into space.

Recovery

With resumption of manned flights dependent on a strenuous redesign and test phase to prevent a recurrence of the tragic loss of Soyuz 1, it was decided to attempt a circumlunar flight of the Zond derivative as some form of celebration for the 50th anniversary of the Soviet Union. On 28 September 1967, a Proton rocket lifted off carrying the unmanned 7K-L1-4 spacecraft, but a failure with the first stage of the launcher sent it crashing to Earth. The spacecraft catapulted out of the inferno, spreading its toxic nitrogen tetroxide and hydrazine propellants across the sky and over the ground.

With rendezvous and docking – a defining capability for the Russians – yet to be demonstrated, flights with the basic 7K-OK Soyuz resumed on 27 October 1967 with the launch of Cosmos 186 (7K-OK-6), the active spacecraft in what was planned as a dual mission to demonstrate that capability. It was joined three days later by 7K-OK-5 (Cosmos 188), reaching orbit just 15 miles (24km) away from Cosmos 186, and rendezvous and docking was achieved on the first orbit after a flurry of automatic thruster firings and orbit changes made fully automatically with the Igla rendezvous and docking system. The only disappointment was that a gap of 3.3in (85mm) in the mechanical interface of the docking ring prevented an electrical connection between the two vehicles.

This was the first time two vehicles had docked together automatically and it demonstrated the technical effectiveness of the Igla system on which so much depended for the Moon-landing programme. Cosmos 186 returned to Earth as planned on 31 October

SOYUZ 7K-OK AND ZOND FLIGHTS 1966–71

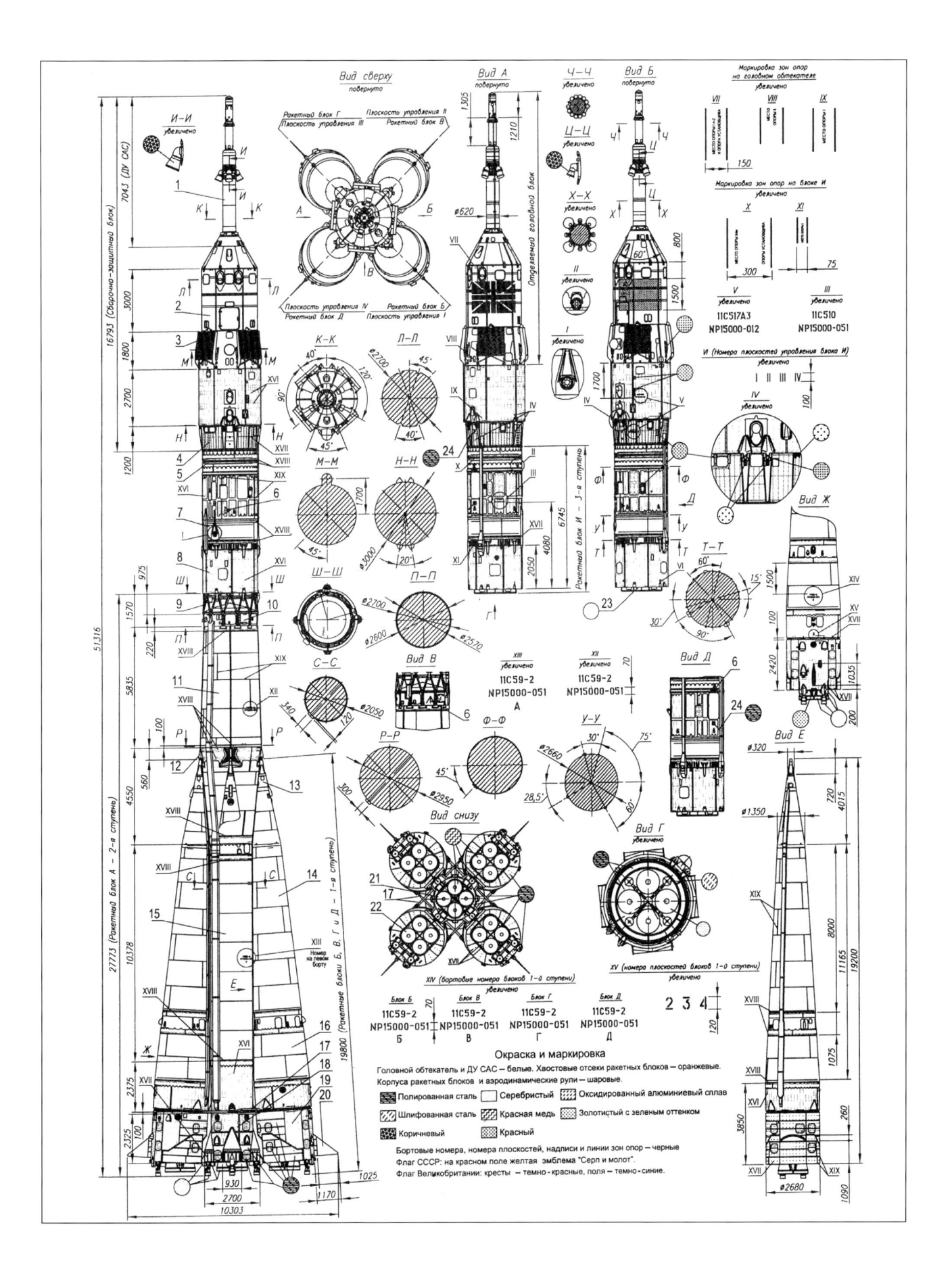

at an elapsed time of 94hr 50min but flew a ballistic trajectory due to a problem with the sensors. Cosmos 188 had the same problem two days later when it returned with a mission time of 25hr 1min but was detonated to prevent it falling outside Soviet territory.

To a national period of celebration and festivity across the Communist world, the 50th anniversary came and went with no spectacular event in space other than the dual flight of Cosmos 186/188. The next attempt in the circumlunar programme came on 22 November when a Proton rocket carrying 7K-L1-5 thundered into the sky but again came crashing to Earth when the second stage failed. The Zond fell to Earth without the aid of a braking rocket to decelerate an otherwise hard landing, which had malfunctioned and fired at altitude.

While modifications to the 7K-OK Soyuz moved ponderously toward a resumption of manned flight, the 7K-L1-06 spacecraft was sent on its way to a circumlunar trajectory by Proton booster on 2 March 1968. The 11,332lb (5,140kg) spacecraft was the first Soyuz to be publicly declared as a Zond mission. Continuing a series that began with the interplanetary probes Zond 1–3 launched in 1964–65, Zond 4 adopted the name of a completely different series to evade suspicion that it was related to the manned programme.

The only flaw in the mission was that an attitude sensor failure modified the re-entry trajectory, causing the spacecraft to fall toward the West African coast and it was blown up to prevent it falling into the hands of the Americans. Returning from the vicinity of the Moon, the thermal load on the heat shield was considerably greater than returning from Earth orbit and the re-entry had been designed to follow a lifting trajectory which was not carried out, the descent module falling to Earth on a ballistic path.

Another blow to Soviet prestige and to Russian pride occurred on 27 March when Yuri Gagarin was killed in an avoidable aircraft accident. This had a deeply demoralising impact on the teams working to restore the Soyuz spacecraft to operational service. Repeated failures with tests of the 7K-OK dragged on through the spring and would drag on through much of that year. Flaws in manufacturing and mistakes caused by pressure to fly condemned almost half of the spacecraft prepared for flight and for automated test.

The push to clear the basic Soyuz for a resumption of manned flights began with the launch of Cosmos 212 on 14 April 1968, in reality 7K-OK-8, as the active partner in a dual flight. It was followed a day later by 7K-OK-7 (Cosmos 213), which reached orbit only 2.5 miles (4km) from Cosmos 212, a stunning achievement. Just 57 minutes after the launch of the target, Cosmos 212 docked with the other spacecraft in another example of impeccable rendezvous technology. The two remained docked for 3hr 50min and separated before returning to Earth, each clocking up about five days in space. Cosmos 212 successfully demonstrated the first guided entry in the Soyuz programme, and although landing safely high winds dragged the descent module about 3.1 miles (5km) from the landing spot!

The next launch in the qualification of the circumlunar Zond variant occurred on 22 April, but the Proton booster suffered a malfunction and blew up shortly after lift-off, although the 7K-L1-7 was recovered. The next attempt, with 7K-L1-8, was aborted when the second stage oxygen tank exploded during flight preparations on 15 July, killing one person. The remainder of the launch vehicle as well as the spacecraft was saved, but it brought a halt to preparations for further flights until this too had been investigated.

News reports of impending flights around the Moon had leaked out to the West and newspapers were already reporting that the Russians were trying to beat the Americans to a flight around the Moon. On 9 November 1967 NASA had successfully conducted a full test of its Saturn V, propelling a test model of the Apollo spacecraft to an altitude of more than 11,200 miles (18,000km) before safely recovering it from the Pacific Ocean. On 4 April 1968 it had followed this with the second flight of a Saturn V, which, although experiencing some technical difficulties, had proven to be a success, clearing the next mission for a manned Apollo flight.

In Russia, the revised flight plan for an already flagging circumlunar programme was being held back by difficulties with the

Proton launcher and not the Soyuz derivative. Nevertheless, tentative plans were made for sending men around the Moon some time in January 1969. In support of that, at least two more automated missions designated in the Zond series would be flown. Meanwhile, the Earth orbit and rendezvous variant was given a final orbital shakedown test with the flight of Cosmos 238. Launched on 28 August 1968, the flight of 7K-OK-9 lasted 94hr 59min and cleared all systems for resumption of manned missions.

Next up was Zond 5 (7K-L1-9), launched by a Proton rocket on 15 September, carrying a host of living things including turtles, mealworms, plant seeds and biological specimens. This was the first time animals and other life forms had been flown around the Moon – passing within 1,212 miles (1,950km) of the surface – and returned for examination on Earth. Although the spacecraft once again experienced failed sensors and problems with automatic attitude control, the flight ended back on Earth safely, splashing down in the Indian Ocean after a flight lasting 6 days 18 hours 24 minutes.

Back to space

The resumption of manned flights following the loss of Soyuz 1 in April 1967 occurred with the launch of Soyuz 3 (7K-OK-10) on 26 October 1968 carrying Georgy Beregovoy. It took place one day after the launch of Soyuz 2 (7K-OK-10), unmanned, to a waiting orbit where Soyuz 3 would rendezvous and dock with it. In attempting to carry out a manual docking with his target, Beregovoy used too much propellant and his manual authority conflicted with the Igla approach radar system, which confounded attempts to bring the two together. Beregovoy had no alternative but to return to Earth, landing after a flight lasting almost four days.

The flight of Soyuz 3 came only three days after the highly successful 11-day mission of Apollo 7, NASA's first manned orbital flight with their flagship Moon craft. Very soon after that mission had been reviewed, news crept out that NASA was considering sending three astronauts into Moon orbit on the next flight, Apollo 8. Tentatively scheduled for December, a

ABOVE The badge for the flight of Soyuz 1, which resulted in the death of Vladimir Komarov on 24 April 1967 on what was to have been the first of a dual flight in which another Soyuz would carry three cosmonauts into space. *(David Baker)*

ABOVE Not before October 1968 did another Soyuz fly with a cosmonaut on board, when Beregovoi spent almost four days frustratingly unable to dock with the unmanned Soyuz 2 spacecraft. *(David Baker)*

spectacularly successful recovery from the redirection of effort after the Apollo pad fire threatened to eclipse any Soviet plan for a mere circumlunar flight – which would at best go around the Moon without even going into orbit.

In a remarkable display of supreme conservatism, the Russians planned for two more L1 flights before considering a manned circumlunar mission in January 1969. In private diaries there were veiled commiserations for the engineers and cosmonauts who had worked for so many years to achieve even these

LEFT The US Air Force was developing its Manned Orbiting Laboratory (MOL), for which Russia was developing a military response with concepts of Soyuz gunships and other defensive mechanisms. In 1966 a test launch was made with a Titan 3C carrying a cylindrical mock-up of the MOL with a Gemini spacecraft on top. *(USAF)*

ABOVE From left to right: V. Khrunov, A.S. Yeliseyev, V.A. Shatalov and B.V. Volynov, the crews of Soyuz 4 and 5, Russia's first successful rendezvous and docking operation, completed in January 1969. *(Novosti)*

modest steps. On November 12, NASA formally announced that Apollo 8, to be launched on 21 December, would fly to the Moon, orbit ten times and return to Earth. Two days after the announcement, the 11,850lb (5,375kg) Zond 6 was launched on a repeat of the preceding mission, carrying the unmanned 7K-L1-12 around the Moon and back.

On returning to Earth a gasket failed, the interior depressurised and the biological samples it was carrying died. The descent module executed a perfectly controlled guided re-entry, first dipping into the atmosphere and then back to the outer fringes to dissipate heat before descending again, imposing little more than 4g deceleration. Instead of landing back close to the launch site at Baikonur, the spacecraft suffered a failure in the parachute deployment mechanism and smashed into the ground.

After this less than totally successful flight, and a consistent history of systems and equipment failures, Mishin acknowledged that there was now little point in short-term planning for a circumlunar flight. The engineers, scientists and workforce were distraught when, in late December 1969, three American astronauts flew the Apollo 8 mission to the Moon, reading an emotional Christmas message to the world, while Soviet hopes lay far from realised. A top priority was to qualify the 7K-OK version of Soyuz and to begin a series of manned rendezvous and docking flights in the coming year.

The flight of Cosmos 238 cleared all remaining issues regarding the next manned flights, and on 14 January 1969 Vladimir Shatalov was launched in the 14,606lb (6,625kg) Soyuz 4 (7K-OK-12) to a waiting orbit where he would be the active vehicle in a dual rendezvous and docking flight. A day later, the 14,715lb (6,585kg) Soyuz 5 (7K-OK-13) spacecraft was launched carrying Boris Volynov, Aleksei Yeliseyev and Yevgeny Khrunov. Launched in shirtsleeves, the crew of both spacecraft achieved the first rendezvous and docking between two manned vehicles 48hr 50min after the launch of Soyuz 4 and 25hr 15min into the flight of Soyuz 5.

Donning Yastreb spacesuits in the orbital module of their spacecraft, Yeliseyev and Khrunov moved to the OM on the still docked Soyuz 5 and reversed the process, joining Shatalov in Soyuz 4. The bulky probe and drogue docking system was a fixture in the forward section of each orbital module and there was no internal access between the two vehicles. The suits worn by these cosmonauts were considerably improved on the one used by Leonov almost four years earlier, with pulley systems and restraint rings to prevent them ballooning and compromising movement. Each man had an independent life-support pack and there were tethers to prevent either floating away. Yeliseyev followed Khrunov, leaving Volynov alone in Soyuz 5.

The two spacecraft remained docked for only 4hr 34min before separating and preparing for re-entry, Soyuz 4 returning with its three cosmonauts after a mission elapsed time of 71hr 20min 47sec and landing as planned after a fully guided re-entry. When Soyuz 5 came home it was a near tragedy, as Volynov reported shortly after the de-orbit burn that the equipment section had failed to separate from the descent module. A similar situation had happened with several Vostok flights, but it was a much more serious matter with the much heavier Soyuz.

With the whole vehicle tumbling end over end, propellant for the DM thrusters was quickly exhausted, and without facing the appropriate direction the heat shield was still covered by the equipment module. As smoke began to appear inside the DM hope for Volynov surviving this flight swiftly evaporated, but in the spacecraft he quickly gathered all the diary records of the flight and stuffed them into a part of the spacecraft he thought might survive the fiery re-entry.

Then he began to quietly read into a recorder his instrument readings and log of activity as all the while the modules, locked together in a wild whirligig ride, spun and rotated in a frenzied attempt to break free. With a sudden *CRUMP* the propellant tanks in the equipment module burst, and the crew hatch deformed inward and then outward under the sudden pressure. With a loud wrenching, grinding of metal and scraping sounds, the equipment module finally broke free, and, seconds from Volynov being burned alive, the descent module righted itself and he flew a bone-crushing ballistic descent.

But as the parachute deployed, the riser lines were twisted and only gradually began to unravel ever so slowly as the surface of the ground came up. With a shattering thump the module struck Earth, smashing off the roots of Volynov's upper teeth inside his jaw. The shock-absorbing effect of the seat was the only thing that prevented him from having severely broken bones.

While the official news report proclaimed a complete success for what was, in reality, a major technical achievement, the engineers were delighted with this unconventional way of returning to Earth. It showed in the most dramatic way possible how well the module had been designed and how even under these conditions it would right itself and achieve stability. It was a very long time before the world got to know how Volynov had come close to death.

While the successful flight of Apollo 8 had eclipsed anything the Russians could do in

BELOW The crew badge for Soyuz 5, an embellishment now traditional for a Russian space programme entering a new age of routine flight operations with a far more versatile spacecraft than had been flown before. *(David Baker)*

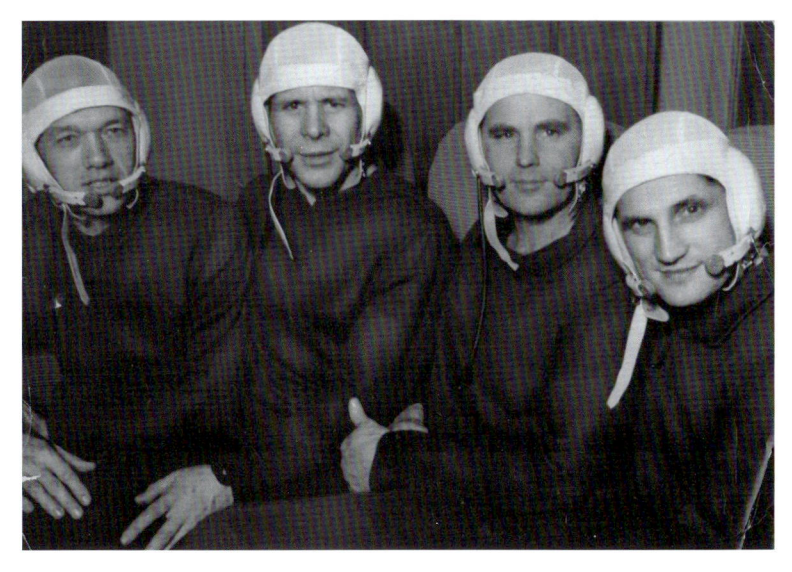

ABOVE Wearing flight suits similar to those they would wear in space, the crews of Soyuz 4 and 5, from left to right Yeliseyev, Khrunov, Shatalov and Volynov. *(Novosti)*

LEFT The crew badge for Soyuz 4 reflects the expectant aspirations of its cosmonaut and support teams for achieving the first link-up between two Russian spacecraft. *(David Baker)*

BELOW Recorded by Soviet tracking stations, the televised docking of Soyuz 4 and 5 was broadcast to millions across the USSR and to a wider audience around the world. *(Novosti)*

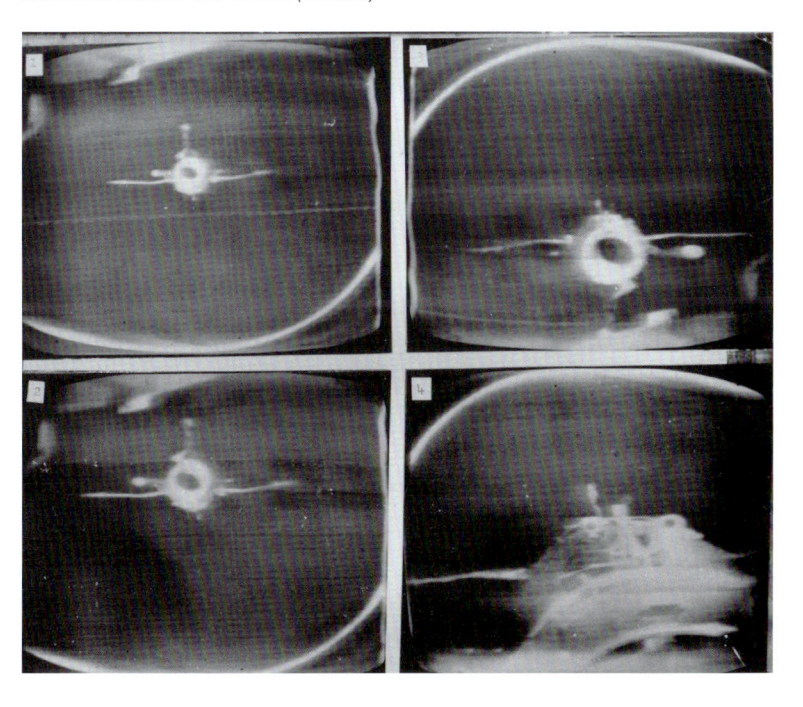

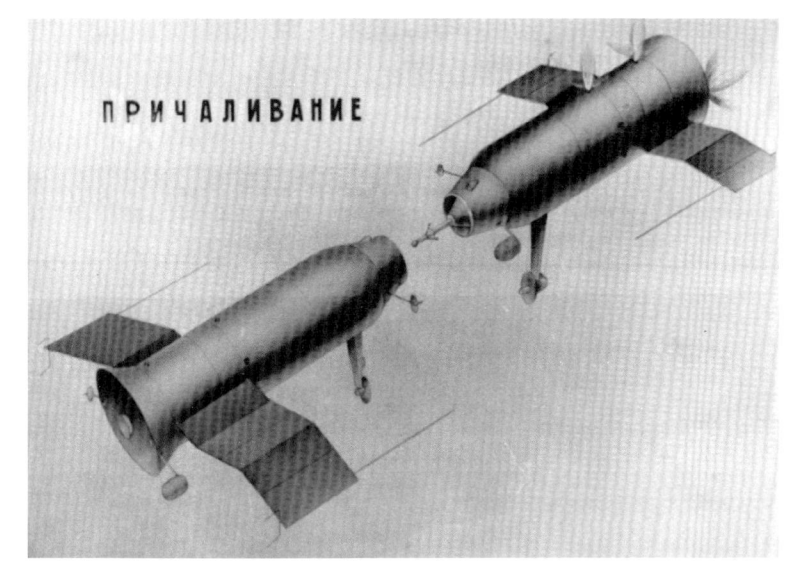

LEFT While the engineers and scientists revelled in the publicity for their achievements, the continual struggle against Soviet censorship was reflected in initial drawings that fail to reveal the precise shape of the re-entry module. *(Novosti)*

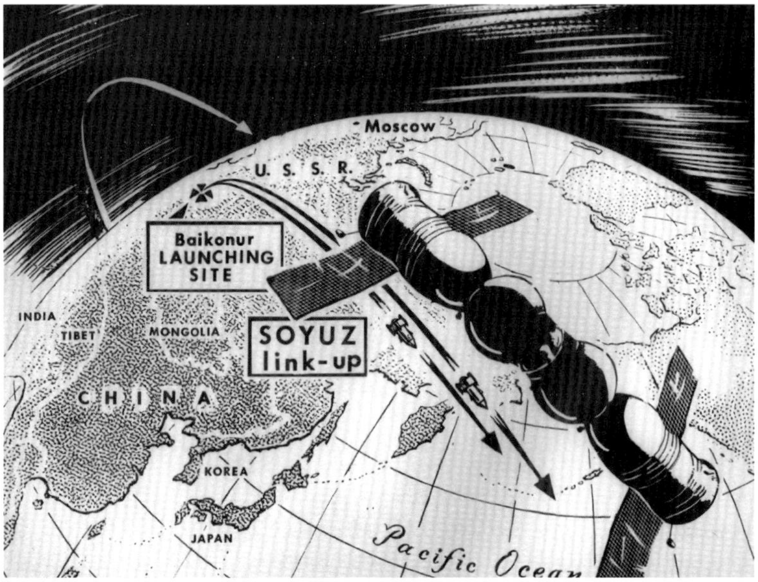

CENTRE This news picture released for publication in the Western press signals a degree of openness and less sensitivity to shrouded detail as it trumpets the success of Russia's space programme, now falling behind American achievements in the year Americans will land on the Moon. *(Novosti)*

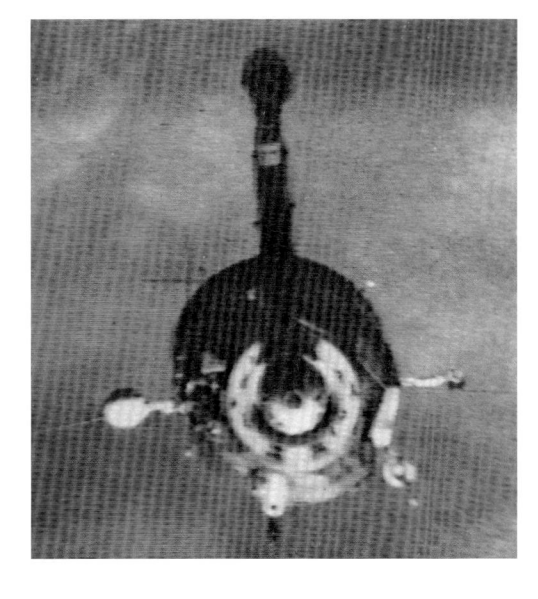

RIGHT A frame grab from the live TV showing the approaching Soyuz spacecraft, probe extended, with radar antenna and periscope protruding. *(Novosti)*

the short term to restore world acclaim in their flagging programme, there was a determined will to transform the circumlunar L1 missions into scientific flights, and to proclaim as much to the world. After all, there had never been any official link between the Zond-Soyuz and numbered Soyuz flights for Earth orbit, rendezvous and docking. On 20 January, 7K-L1-13 was launched atop another Proton rocket but a failure in the second stage ended the flight 501 seconds after lift-off, the fourth failure of a Proton in nine launches.

Just a month later, on 21 February 1969, the giant N1 rocket designed to carry a complex group of modules, a variant of the Soyuz spacecraft (7K-L1S), and a lunar lander on a test flight, ascended from the Baikonur cosmodrome with a lift-off thrust of 10.12 million lb (45,017kN). It was Russia's equivalent of the Saturn V, more powerful but with slightly less payload capability than the American rocket. The flight was to have vindicated Soviet plans for a Moon landing within the next year or two. That hope was dashed dramatically when a fire broke out in one of the first-stage engines less than 70 seconds after launch, bringing the whole assembly crashing to the ground.

Just two weeks later NASA flew its much-vaunted Lunar Module, along with the Apollo mother ship, on the Apollo 9 mission in Earth orbit launched by the fourth successful Saturn V in a row. Demonstrating separation of the two manned vehicles, free flight up to a distance of 115 miles (185km) apart, rendezvous and docking, it cleared the way for NASA to send the next mission, in May,

ABOVE Great detail was allowed by the censors when they released full-scale mock-ups of Soyuz 4 and 5 in the docked configuration, allowing visitors to exhibitions to scrutinise in detail the intricacies of the new spacecraft. *(Novosti)*

ABOVE Without a camera on hand at a distance, this is as close a rendition of how the two vehicles looked as it was possible to achieve. Nevertheless, the sheer scale of the docked vehicles across a total length of almost 50ft (15m) is impressive. *(Novosti)*

on a full dress rehearsal of a Moon landing short of descending to the surface from a low lunar orbit. Two months later, on 20 July 1969, Apollo 11 put Armstrong and Aldrin on the Moon, accomplishing the goal set by President Kennedy eight years and two months earlier.

It is ironic that just as the Americans were stealing the limelight, Russian engineers were about to send up the first Zond spacecraft equipped to carry cosmonauts. Built as the last configured for automated flight, 7K-L1-11 had been reconfigured for piloted operations to fully test the vehicle before sending people. Designated Zond 7 when launched on 8 August, it flew a near perfect circumlunar mission, packed with living things for scientific analysis of radiation levels and extended weightlessness, returning to land on Soviet territory only 31 miles (50km) from the designated touchdown spot.

The aftermath of the American success and the total inability of the Soviet propaganda machine to extract anything worthwhile from a manned circumlunar flight doomed the Zond versions of Soyuz to a few remaining unmanned scientific flights investigating the reaction of living structures to the space environment. Better, said the policy makers, to walk away from the Moon and disassociate the Soviet programme from any suggestion of running behind the Americans.

It is difficult to understand, in these circumstances, how the effort to land men on the surface of the Moon continued to extract

BELOW The bulk of the early docking probe and drogue assembly is evident from close examination of the hardware, illustrated here in scale, which also shows the transfer of spacewalkers Khrunov and Yeliseyev from Soyuz 4 to Soyuz 5 where they joined Shatalov in the descent module. *(David Baker)*

RIGHT The Soviet publicity machine does its best to convey the scale and capability of its new spacecraft, recovering from a lapse in manned flights occasioned by the death of Komarov in Soyuz 1. *(Novosti)*

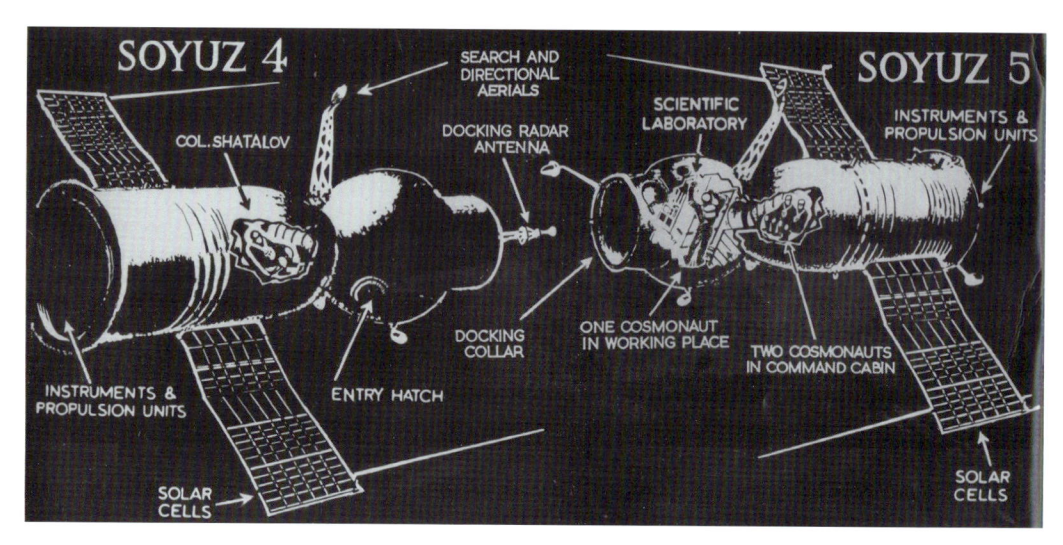

resources in a programme that was already faltering and failing to deliver. On 3 July 1969, less than two weeks before Apollo 11 left for the Moon, the second N1 was launched, but it too failed shortly after lift-off. The N1 programme – doomed by early indecision and bitter in-fighting between design bureaus which lacked the central leadership that could have been expected from the Soviet system – never achieved a successful flight. The last two N1s launched, on 28 June 1971 and 23 November 1972, also failed to reach space.

As a footnote to the fallout from these failed objectives, early in 1969 a rushed effort was made to try and get to the surface of the Moon with an unmanned lander and return to Earth with samples before Apollo 11. Launched on 13 July 1969, Luna 15 eventually crashed into the surface and eliminated that possibility. Nevertheless, the Luna programme did eventually achieve creditable results of great scientific value, returning small quantities of samples from the surface and deploying roving vehicles which continued to operate for several years.

During the first half of 1969 the Defence Ministry took a hand in re-orientating the Soviet space programme. There would be

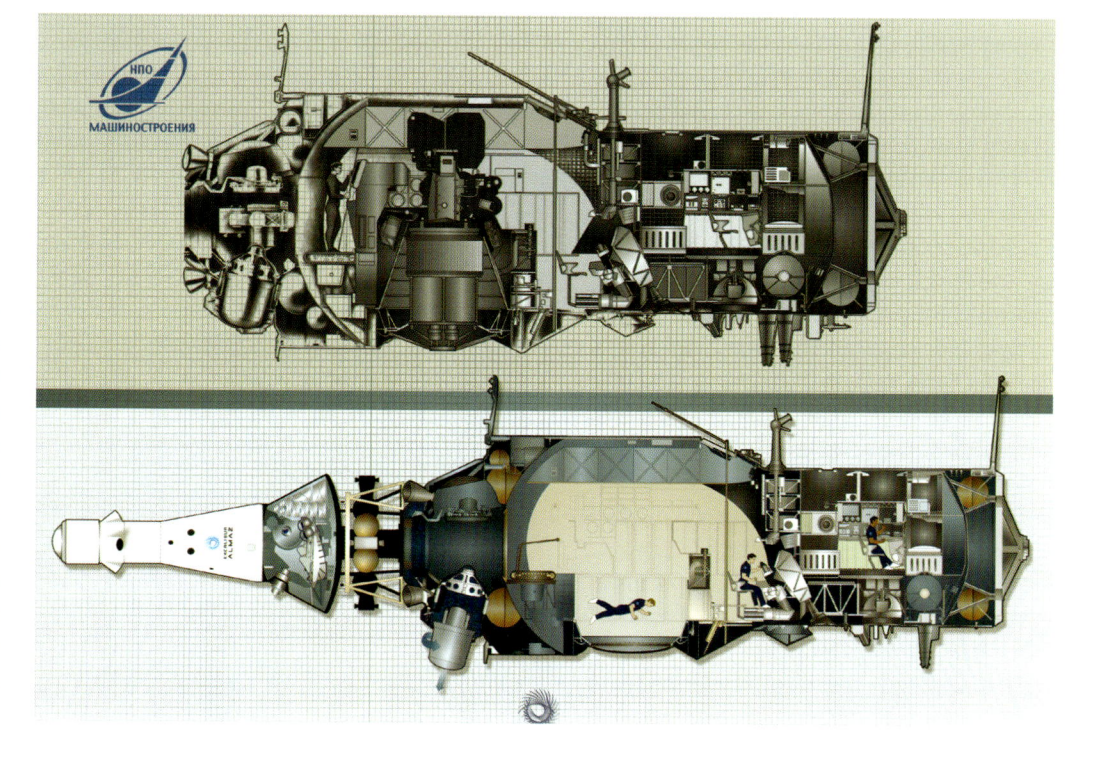

RIGHT The development of the Almaz military space station was becoming a priority with the Russian military as they chafed at continued investment in publicity-seeking space activity at the cost of their own space programmes. *(David Baker)*

no manned circumlunar missions, the N1/L1 programme would be placed in second priority to development of a space station and full development of the basic Soyuz would continue, effectively seeing the future for this vehicle as a ferry to a permanently manned orbital scientific research facility. It would be a very long time before the world would learn about the deep trauma experienced by Russian space scientists and engineers in the aftermath of the Apollo success.

To achieve some level of dynamic publicity, it was decided to fly a triple-Soyuz mission with rendezvous and docking involved, which fed directly into the increasing support for a Soviet space station programme. Several technical developments had been made, including improvements to the Igla rendezvous and docking system and its potential replacement by a system known as Kontakt. This required a manual approach and docking with a triple-pronged grappling system on the active vehicle engaging with a hexagonal grid with 108 receptacles on the passive spacecraft.

Kontakt had been developed in the early 1960s and was one of the two finalists in a design selection review that chose the Igla system. In some ways more sophisticated than Igla, it was selected for the Soyuz L3 variant which was to have been used as the mother ship on lunar landing flights but got sidelined due to its complexity. Involving a different rendezvous radar, it would eventually be used on later Earth-orbiting Soyuz flights. Along with other technical upgrades, the Kontakt systems had been proposed for the triple-docking attempt.

The triple flight began on 11 October 1969 with the launch of cosmonauts Georgy Shonin and Valery Kubasov aboard Soyuz 6 (7K-OK-14), followed a day later with the launch of Soyuz 7 (7K-OK-15) carrying Anatoly Filipchenko, Vladislav Volkov and Viktor Gorbatko. Soyuz 8 (7K-OK-16) followed a day after that with Vladimir Shatalov and Aleksei Yeliseyev. With three spacecraft and seven cosmonauts in orbit at the same time, two new records had been set. On 14 October all three Soyuz were within visual distance of each other, communicating both with the ground and each other.

Several science tasks had been planned into this triple mission and there was full and effective news coverage explaining a range of tasks being carried out, including experiments into material processing whereby weightlessness allows the preparation of fluids and molten metals in a different molecular matrix to that possible in a 1g environment. Earth photography, observation of missile test flights from the USSR and a range of other experiments occupied media interest, but the real purpose of the mission was running into trouble when a succession of minor technical glitches prevented Soyuz 7 and 8 docking as planned, with Soyuz 6 standing off some distance, its crew taking pictures of the event.

The three Soyuz spacecraft returned to Earth on three successive days, in numerical order beginning with Soyuz 6 on 16 October. Each landed safely and without complications, improvements having been made to the separation systems between the various modules to prevent a recurrence of the hang-

BELOW To satisfy an urgent need to put cosmonauts around the Moon before the Americans, the L1 Zond derivative was successfully sent on circumlunar missions, the spacecraft itself essentially a basic Soyuz but without the orbital module. *(David Baker)*

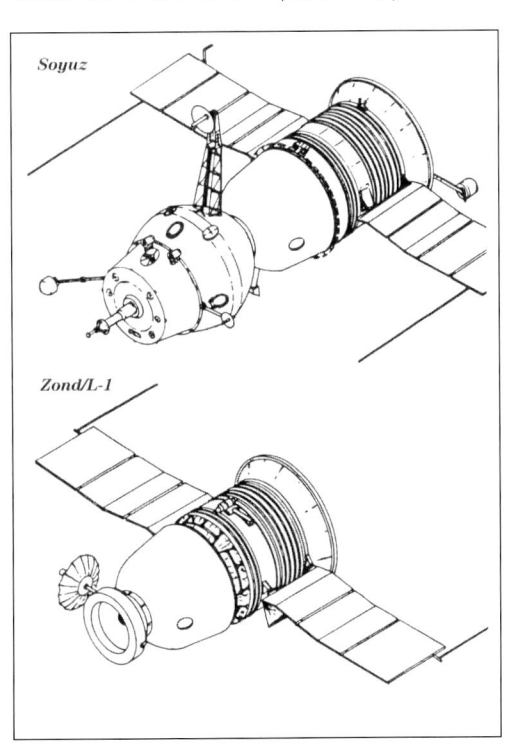

BELOW The Proton rocket was specifically designed for a wide variety of missions, but the one which saw its most prolific application in the late 1960s was the launch of Zond spacecraft. *(David Baker)*

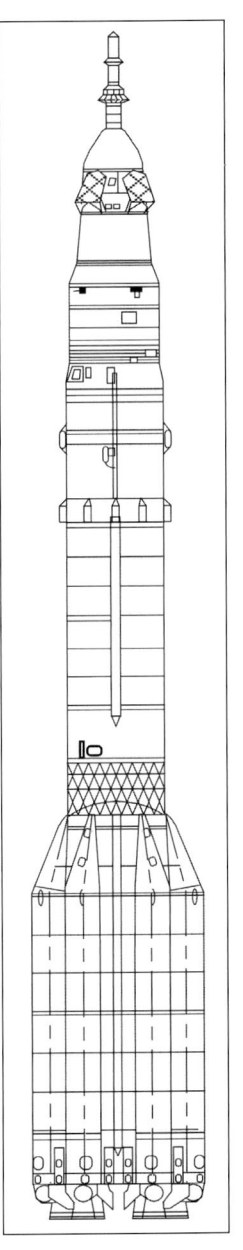

RIGHT Zond 5 splashes down in the Indian Ocean in September 1968 just three months before NASA will send Apollo 8 to orbit the Moon ten times at Christmas. *(Novosti)*

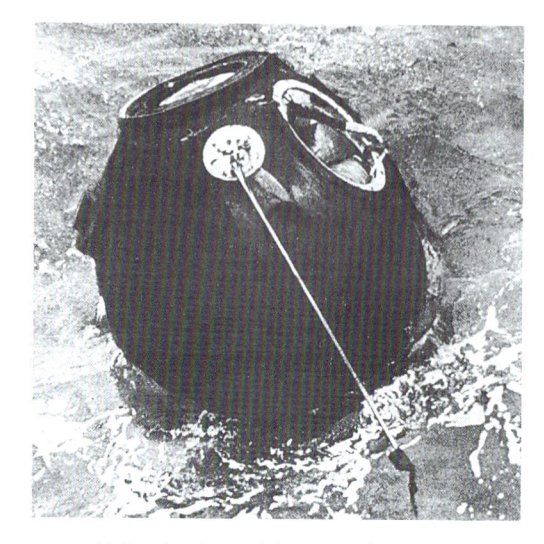

up on Volkov's descent ten months earlier. The failure to dock was masked by abject denials that such an operation was planned, but the diaries of now deceased engineers and managers allude to the frustration over irritating problems with the Igla system and other equipment on board.

BELOW On 21 July 1969, Neil Armstrong and Edwin 'Buzz' Aldrin (seen here) set foot on the Moon and deploy three simple instruments, forerunners of five more landings and many more instruments set down on the lunar surface by the end of 1972. *(NASA)*

A change of direction

Just four days after Soyuz 8 landed, on 22 October Leonid Brezhnev, First Secretary of the Communist Party, gave a speech in which he defined with a clarity that had never been heard before, the future objectives of the Soviet space programme. In it he asked Soviet citizenry to cast their eyes toward space as a place to improve the quality of life, a place to seek out improvements to the national economy and to the livelihoods of people across Russia. Extolling the virtues of conducting science in space, of making new discoveries and working with new materials capable of being manufactured in weightlessness, he placed the Soviet space programme on a higher intellectual plane than the politically motivated 'flag and firsts' ambitions of the United States.

Of course, this was politicking from a consummate politician and did nothing to assuage the deep feeling of disappointment among space scientists and engineers that so many of the grand ideals of the early years, when all seemed so fresh and achievable, had been dashed by mismanagement, indecision and a string of seemingly insurmountable technical hurdles. It was a change in emphasis forged, like so many decisions made by Soviet politicians, in reaction to developments in America.

With falling budgets and an end to Apollo Moon missions in sight, NASA was turning to the use of redundant hardware for a series of long-duration flights aboard a space station called Skylab, which would be launched in 1972 or 1973. Three astronauts would spend 28 days aboard Skylab and return, followed shortly thereafter by a flight lasting 56 days and a final repeat flight of the same duration with another crew. Beyond that, America was talking about a reusable transportation system, the Space Shuttle, and bigger space stations and space bases accommodating up to 50 people.

But America was also talking to Russia, with

BELOW The jubilation which Russia had long hoped would be its reward for having pioneered a path into space was given to three NASA astronauts in this New York parade by grateful Americans who saw the fulfilment of President Kennedy's dream. *(NASA)*

offers of uniting in a joint flight to involve Apollo and Soyuz spacecraft working together and sealing a new era of friendship and cooperation. There was a new president in the White House. Richard Nixon was keener than most to wipe his hands of the Kennedy goal, bury Apollo and look more to how his foreign policy initiatives could be enhanced, and not directed by, a competitive confrontation with the USSR. Slowly, and quietly, the two sides began to talk about a possible deal that would exploit the last existing Apollo spacecraft and the very new Soyuz spacecraft.

But that was some way off. For the second half of 1969 a complex and interwoven series of discussions, political shuffling and decision-making resulted in a commitment to a 'Long Duration Orbital Station', bearing the acronym DOS in Russian. Mishin would have responsibility for the DOS and Chelomei was to turn over his Almaz hardware, still languishing in development tests, to help him accelerate fulfilment of this objective. The new project would be designated DOS-7K, because Soyuz would now be an integral component of that station's ability to remain in space through ferry flights and replenishment missions.

Because of these decisions, Russia turned to using Soyuz in a new direction, and that was epitomised by the next flight when Soyuz 9 (7K-OK-17) was launched on 2 July 1970 with Andrian Nikoleyev, returning to space after his Vostok 3 flight, and Vitaly Sevastyanov. The way the flight had been designed reflected the new emphasis on long duration stays in space, and for that the crew had the benefit of a small heater in the orbital module for warming up food and drinks. They had a few comfort items and a routine previously unknown to cosmonauts, cycling into a timetable which provided plenty of time for working the 50 or so experiments carried on board.

With a daily intake of 2,600 calories, as the mission progressed into its second week there were signs of fatigue and listlessness, the crew reducing their liquid intake to a litre a day and consuming only 17 litres of oxygen per day. They did have an exercise device to maintain muscle tone and load the cardiovascular system, but as the flight extended into its third week they were pushing into unknown territory on flight duration and the medical effects of sustained weightlessness. Despite consideration of an extension, Soyuz 9 ended with a landing on 19 June, a total mission time of 17 days 16 hours 58 minutes 55 seconds.

The 7K-OK Soyuz had been the design basis upon which Earth orbit rendezvous and docking could be evaluated, circumlunar flights could be conducted with the 7K-L1 (Zond) variant, and manned lunar landings supported with the 7K-L3. With the latter two cancelled and in serious doubt respectively, the 7K-OK would evolve into the third variant for a role that would define it for the next several decades – that of a ferry vehicle to and from Earth-orbiting space stations.

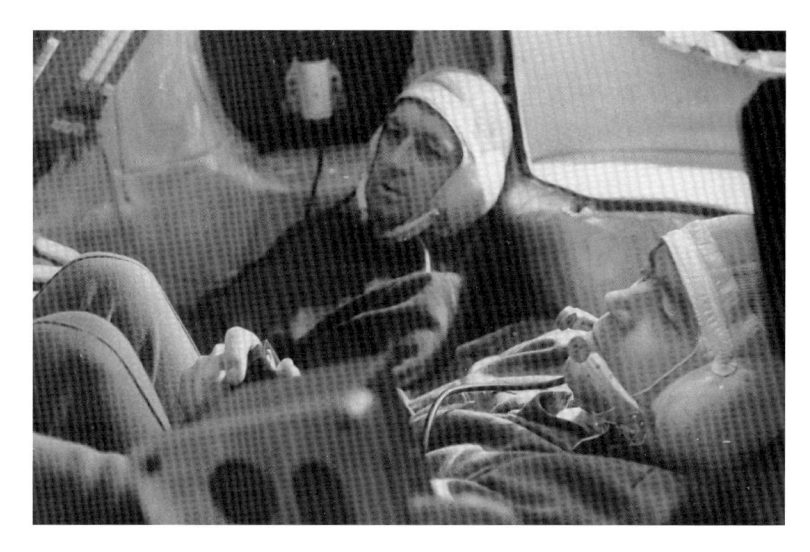

ABOVE The Moon landings had a profound impact on the Russian space programme, transforming an expeditionary zeal into a desire to use space for applications that could directly benefit people on Earth. It was a message echoed by America. In July 1970 Soyuz 9 carried Andrian Nikolayev (right) and Vitaly Sevastyanov on a mission lasting almost 18 days, seizing the duration record set by Americans Borman and Lovell in December 1965. *(Novosti)*

BELOW The end of the beginning as Soyuz 9 returns to Earth and signals a new direction for Soviet space ventures, a commitment to orbital laboratories, space stations and a permanent human presence in orbit that endures to this day. *(Novosti)*

Feature B

Rendezvous and docking

The history of rendezvous has been a long and evolving story, with early evaluations of different techniques beginning with the Gemini VI/VII mission in December 1965. But the study of rendezvous of one object and another in space goes back long before that. In Russia the eminent writings of space prophets, mathematicians and engineers resulted in a set of principles that were excellent in theory but operationally unproven. Most eminent, and revered today around the world, was the Russian teacher and theoretician Konstantin Tsiolkovsky, born in 1857.

Tsiolkovsky inspired a generation of rocket engineers in the early years of the 20th century, and while confining himself to the principles of rocketry and space travel his writings foretold of a period in which humans would colonise the solar system and eventually all of space. His writings did not include the mathematical problems of rendezvous between objects in space, but Russian theorists wrestled with the concept for several years before engineers produced the first Soviet spacecraft capable of changing orbit to achieve rendezvous and docking. In Soyuz, Soviet space programmes could grow and advance beyond the early stages where Vostok and Voskhod spacecraft, unable to change orbit, were denied the essential tools for bringing two objects together in space.

In the United States, serious study of rendezvous techniques began shortly after the first satellites were launched, and in 1959 the US Air Force proposed a vehicle which could hunt down Soviet satellites and either disable or destroy them. Known as SAINT, and later as Program 621A, it stimulated theoretical analysis of just how that could be achieved. In 1959 Edwin 'Buzz' Aldrin wrote his PhD thesis on orbital rendezvous and laid the foundation for what became an essential prerequisite for a wide variety of missions involving humans in space.

The US hunter-killer satellite was never developed but it laid the foundation for extensive tests with a wide variety of rendezvous and docking techniques during the Gemini programme. Gemini was NASA's second-generation spacecraft, a two-man vehicle capable of remaining in space for two weeks, and the first practical demonstration of how to change the geometry of the orbit when relating to the fixed motion of another vehicle. By the time Gemini began carrying humans into space in 1965 the technique was crucial for the Apollo mission to the Moon.

Proclaimed in May 1961, NASA's Moon goal for astronauts was bequeathed by a President who until that date had been almost totally dismissive of space. It was only the spectacular flight of Yuri Gagarin in April 1961 that propelled the American political machine into a self-imposed race with the Soviet Union. Only when extensive evaluation of potential ways to reach the surface of the Moon pointed to the Lunar Orbit Rendezvous mode as being the most likely to achieve an early result, did rendezvous and docking become an integral part of the race with Russia.

By using a separate Lunar Module to descend to the surface of the Moon, the essential feature of the safe return of the crew was to lift off from the surface and rendezvous in lunar orbit with the

BELOW The docking system on the modern Soyuz is a development of several earlier design concepts, each of which was relevant to different roles played by Soyuz. In this definitive arrangement on Soyuz TMA, after docking the probe can be swung down and out of the way of the docking interface, opening a passage for crew members. *(NASA)*

Apollo mother ship. Rendezvous and docking was only a theoretical idea with several potential problems, not least the very high relative velocities of spacecraft in different orbits and the potential for catastrophic collision. The science of orbital rendezvous is replete with counter-intuitive realities unfamiliar to Earth-dwellers used to two-dimensional manoeuvring.

The physics of celestial bodies determine the way orbital flight is achieved and this plays into the dilemma of how to bring two objects together in space. The most important aspect for any rendezvous is to ensure the target vehicle is in a fixed orbit relative to the body around which it is circling. It helps greatly if the chase vehicle is launched into the same orbital plane as the target. That is, if the inclinations of the two orbits are the same, or coincident, the function of the chase vehicle is to enter an orbit which is lower than, and behind, the target. Plane changes are very costly on propellant; for a vehicle of the mass of Soyuz a plane change of 1° would require a ΔV of 290mph (466kph).

The orbital inclination of the International Space Station is 51.6° to the Earth's equator and it is to this orbital inclination that the Soyuz chase vehicle is launched. However, because orbits are fixed to the Earth's centre of gravity, and the Earth is itself rotating on its polar axis around that centre, there is only one time each day when the launch site rotates to lie under the orbital plane of the target. This is the time to launch and for Soyuz to enter an elliptical path lower than, and behind, the ISS.

Orbital velocity is set by the force of gravity at specific heights above the Earth. Because the force of gravity declines on the inverse square law, the higher the object is above the surface the slower it travels around the Earth. In effect, objects are always under the influence of gravity but 'weightless' because they are continuously falling toward the Earth, but never reaching the surface because as they move forward the curvature of the Earth keeps receding. This is what defines orbital velocity. Because the chase vehicle is in a lower orbit it will travel faster round the Earth than the target vehicle in a higher orbit. This is why it must be placed in orbit behind the target, slowly catching up.

The angle between the two is known as the phase angle, and this will gradually close as the

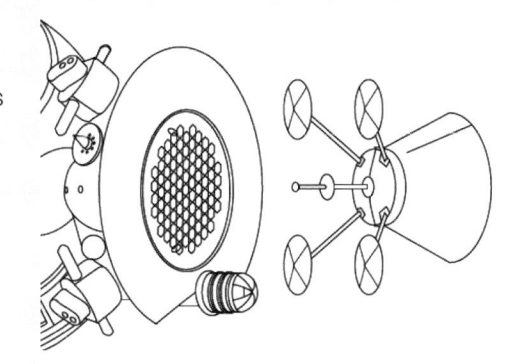

LEFT The primitive Kontakt docking system was developed in the early stages of Soyuz adaptation for the role of Earth-orbiting laboratory, when the connection of two vehicles in tandem was considered a priority. Here the system used a series of probes engaging with multiple receptors. *(David Baker)*

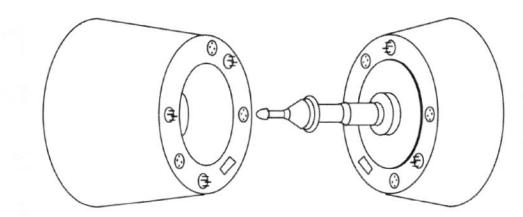

LEFT The 7K-OK Soyuz docking system used for the initial demonstration docking operation involving Soyuz 3, 4 and 5 in 1969 used a probe and fixed drogue system which gave no opportunity for intra-vehicular transfer, necessitating a spacewalk to move from one vehicle to another. *(David Baker)*

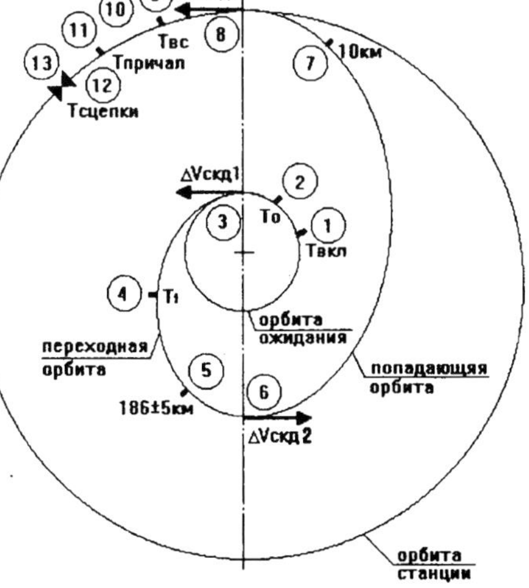

LEFT The rendezvous process for Soyuz on a mission to the International Space Station begins shortly after orbital injection (1), followed by determination of the relative phasing angle from onboard readouts and ground tracking parameters (2). From a circular orbit, the first phasing burn is performed (3), after which the Kurs equipment is switched on (4) and a determination made of the range and the range-rate of the spacecraft (5). At apogee (6) the elliptical orbit thus obtained at the first burn is translated into a co-elliptic manoeuvre by a further engine burn, and a Kurs test is made at a distance of 6 miles (10km) (7). A further terminal phase burn is made (8) and a distance established (9) within a range of 1,310ft (400m) and where the closing rate is less than 0.3°/sec of phasing angle. Fly-around and station-keeping begins (11), followed by the final close approach phase (12) and docking (13). Recently, this rendezvous sequence has been considerably abbreviated and docking is achieved within a few hours. *(Energia)*

ABOVE Developed for the Apollo–Soyuz Test Project (ASTP), the APAS-75 system evolved from the need for an androgynous design, unlike earlier US and Russian docking systems which had male probe and female drogue components. The word APAS derives from a Russian phrase combining 'complex' and 'mating'. APAS-75 was used on three test flights, Cosmos 637, 672 and Soyuz 16, and on the manned Soyuz 19 spacecraft employed for the actual mission. *(David Baker)*

BELOW APAS-89 was developed by the Russians when they envisaged operations with the Mir-2 space station and the reusable Buran shuttle vehicle. Neither was in fact used operationally but APAS-89 was used on one of the modules for the Mir station that did see extensive use. This design was first flown on Soyuz TM-16 in 1993 and was also used by the Shuttle as a docking interface with Mir. *(David Baker)*

two converge on the same position over the surface of the Earth. If the chase vehicle were to remain in this lower orbit it would overtake the target and move ahead, but remain significantly below and never rendezvous with it. And, because the higher the vehicle is above the Earth the slower it moves, as it is placed into an elliptical orbit with the high point closer to the altitude of the target, it not only moves up toward the target but slows down too. Then, as the chase vehicle follows a descending path toward the low point of its orbit, it speeds up and closes the phase angle at a faster rate, before once again climbing, slowing down, and closing the distance and the phase angle to the target. This is known as a co-elliptic path.

In most rendezvous techniques, the initial elliptical orbit is circularised at the high point, or apogee, by slowing the speed of the chase vehicle, raising the low point, or perigee, half an orbit later. This is known as a concentric orbit. With the orbit of the chase vehicle now closer to the orbit of the target, the relative velocity between the two will be less than it was in the co-elliptic path, the two in relatively stable orbits, with the chase vehicle now closing the phase angle at a slower rate. The next phase is to repeat the operation again, transferring the chase vehicle to a new elliptical orbit, this time with apogee at the same height as the target.

When the apogee of the chase vehicle matches that of the target the chase vehicle should be either behind or in front. A complex sequence of manoeuvres begins so as to achieve station-keeping, where the two are virtually motionless in space. But even here,

BELOW The active component of the APAS-89 androgynous docking system, defined as such by the extended ring clutch. *(David Baker)*

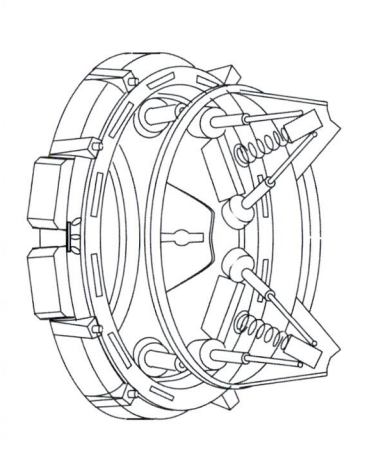

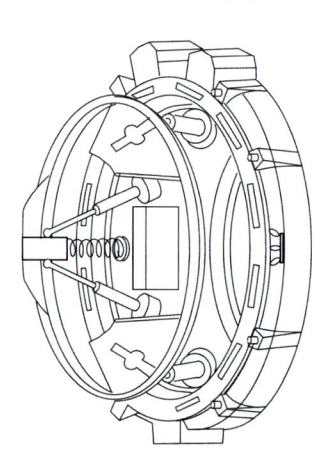

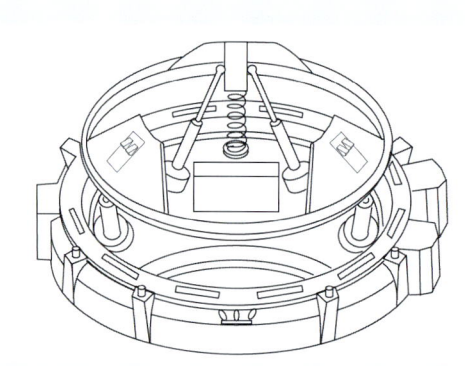

small manoeuvres with the thrusters upset the stability of the two, which are still in their separate orbits. If the chase vehicle speeds up it will rise above the target. If it reduces its forward motion by thrusting in the direction of travel in a braking action, it will start to sink below the target. Eventually these fluctuations will null and the two vehicles can dock together.

The first rendezvous mission was Gemini VII in December 1965. Shortly after reaching orbit the crew practised the technique of 'station-keeping', using thrusters to maintain their spacecraft in close proximity with the second stage of their Titan II launch vehicle. In a series of errors accumulated through lack of experience and the need to develop new skills and techniques, the spacecraft came perilously close to colliding with the stage.

After using thrusters to move away from the stage on separation, the crew turned their spacecraft around to visually track the inert structure. With the Sun behind the stage it blinded the crew, so they pulsed the thrusters in an out-of-plane manoeuvre to shift the Sun angle. But the rocket stage was venting excess propellants to relieve pressure in the tanks, a normal process after reaching orbit, and it too had a displacing effect on its orbit. Unfortunately, the attitude of the Gemini spacecraft relative to the Titan stage was different to the values that had been worked out before the flight. Conforming to a set of pre-planned values, when the spacecraft was not pointing precisely in the direction expected, it compounded relatively minor errors. The combination of these factors had the effect of periodically bringing the two structures together as they criss-crossed orbital paths. Had the Gemini VII spacecraft collided with the Titan stage the relative velocities were sufficiently high to have destroyed the manned spacecraft and killed the crew. None of this complexity, unravelled after the flight, was known during the mission itself.

The flight turned into an outstanding success when Gemini VI, launched several days into the two-week-long mission, rendezvoused with Gemini VII and provided photo opportunities that put pictures on the front pages of newspapers around the world. The first demonstration that two spacecraft could find each other in space, match orbits and manoeuvre around each other under complete control. For the Americans, the path to the Moon was open, and for the Russians it was a reminder that their Soyuz spacecraft, itself capable of rendezvous and docking, was long overdue.

On the next American flight, in March 1966, Gemini VIII successfully docked to an unmanned target vehicle after chasing it down in space, and during the next four missions in the Gemini series a wide range of different rendezvous and station-keeping techniques were rehearsed, demonstrated and evaluated. When the Apollo missions started in late 1968

ABOVE The APAS-89 docking unit playing the passive, retracted role as mounted to the Kristall module on Mir to which the NASA Shuttle would dock. *(NASA)*

BELOW STS-76 docking with Mir, showing the active side of the APAS-89 unit. *(NASA)*

BELOW AND BELOW RIGHT The APAS-95 is a refinement of the APAS-89 unit but with provision for an active capture ring which extends forward and mates with the passive ring, allowing alignment, retraction and capture, with 12 hooks providing a firm connection and an airtight seal. The active side is shown raised, the passive side viewed head-on during approach. *(NASA)*

the procedures essential for getting to the Moon were in hand, and in 1969 three Moon missions, of which two went down to the surface, demonstrated that a Lunar Module could search out and find the Apollo spacecraft in lunar orbit.

The ability for Russian spacecraft to conduct orbital rendezvous and docking required a capability for orbital manoeuvres using propulsion systems like Gemini and Apollo. The Vostok and Voskhod spacecraft only had attitude control, essentially rotation around any of three axes – pitch, roll and yaw – but not an ability to conduct orbit changing, or 'translation' manoeuvres. The Soviet propaganda machine sought to compensate by flying pairs of Vostok spacecraft launched directly into close proximity, thus achieving, in the public view, rendezvous of two vehicles in space.

The first Russian rendezvous operation, conducted by one spacecraft independently approaching another, was the dual flight of Cosmos 186/188 in October 1967, described elsewhere in its historical context as a reaction to the loss of Soyuz 1 and its occupant, Vladimir Komarov, on 24 April 1967. It was the first of many automated rendezvous and docking operations that reflected a degree of control over the mission that was reminiscent of the original design objectives in Vostok. It was repeated with the unmanned flights of Cosmos 212/213 in April 1968, preceding the first rendezvous and docking with cosmonauts on board when Soyuz 4 and 5 were launched a day apart in January 1969.

This dual flight achieved the first docking of two manned space vehicles in orbit, preceding the flight of NASA's Apollo 9 by two months, when a manned Lunar Module docked with a manned Apollo spacecraft in Earth orbit clearing the way for a reconnaissance of Moon landing sites in May and the first manned landing in July. But the means of achieving a successful rendezvous and docking were very different to those adopted by the Americans. Whereas the NASA system was an entirely manual operation involving directly piloted techniques from the astronauts, the Russian system depended upon an automated system known as Igla, which is Russian for 'Needle'.

The system involved a complex of five types of antenna arranged on passive and active vehicles. The passive target carried a radio system that would react as a radio beacon for the active vehicle to home in on and as a transponder for the distance to the active vehicle (range) and the rate at which it was closing in (range rate). The passive vehicle carried an omni-directional antenna with a cardioids hemispherical pattern of a conical helix design with fixed antennae located on two opposing booms on the passive vehicle. Both passive and active vehicles carried two horizontally opposed reception antennae to provide information on relative orientation of the two vehicles.

The active vehicle also had a gimballed dish antenna acting as a transceiver to track the motion of the passive vehicle, this rate being line-of-sight indicating potential miss distance during the approach. This fed directly into the control computer for the main propulsion system for appropriate orientation and firing commands. The passive spacecraft had a fixed dish antenna for transmitting ranging signals to the active vehicle.

The homing phase began when the passive spacecraft transmitted a carrier wave beacon signal through the two transmitting antennae, acquired by one of the rotating receiver antennae on the active vehicles. The active spacecraft would be in a slowly rotating mode to allow the fixed antenna to scan ahead for signals. The first rendezvous of Cosmos 186/188 showed the effectiveness of the system when it demonstrated acquisition at a range of 15 miles (24km). Once acquired by the passive vehicle, the active spacecraft would transmit an interrogation signal through its narrow-beam antenna, which would be locked on to the passive vehicle. The omni-directional beacon on the passive vehicle would then be switched off as the fixed narrow-beam antenna pointed at the active spacecraft.

Tone-ranging was used to determine the distance between the two spacecraft, with an audio frequency of 800Hz modulated on the outgoing carrier and returned by the active transponder on the passive vehicle. Range in the active vehicle was obtained by comparing the different phases of the transmitted and received audio tones, with a voting logic between several different frequency modulations to remove inaccuracies. The passive spacecraft transmitted at 3.2GHz and received at 3.3GHz, with the active vehicle transmitting at 3.3GHz and receiving at 3.2GHz.

The smoothed rate signals indicated the rate of rotation, used as a measure of the coordinates to the passive vehicle. In the event that the line-of-sight was zero the two vehicles would miss each other, but at very close distance this would ensure final approach with the range rate signal as a

ABOVE Progress M-13M draws closer to dock with the International Space Station using the Kurs system and the probe and drogue concept. *(NASA)*

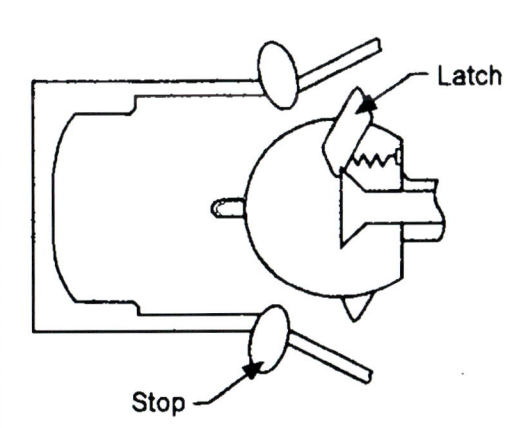

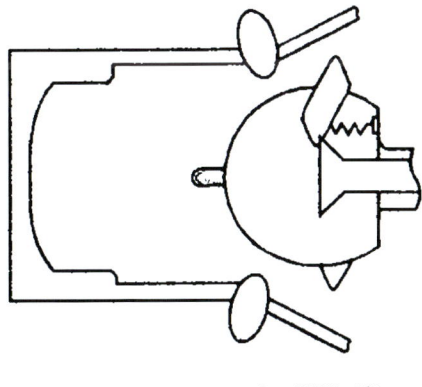

LEFT The probe latching catches for hard contact and docking, a process that also completes an electrical circuit and triggers an advisory on contact. *(David Baker)*

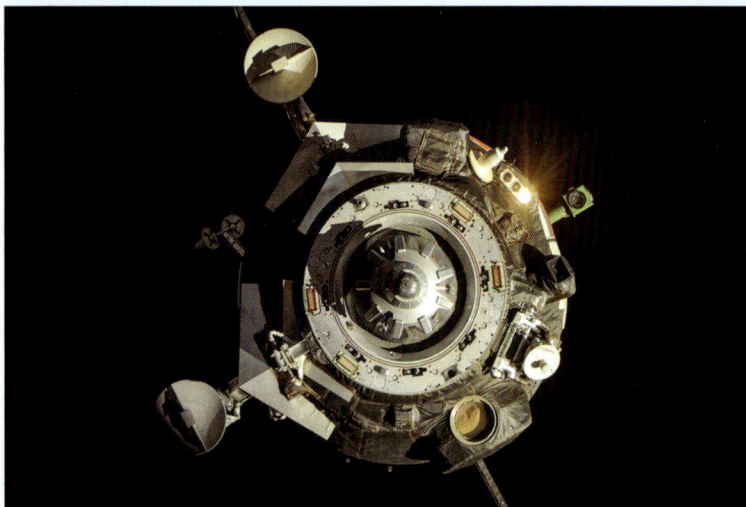

LEFT Soyuz TMA-09M, the latest version of this spacecraft, backs away from the International Space Station on 10 November 2013, displaying the probe, its latches retracted. *(NASA)*

CENTRE TMA-09M fires its thrusters to begin the process of separating from the ISS before returning to Earth. *(NASA)*

function of range providing the value of the diminishing closing distance.

The Igla system would bring the two spacecraft together at a distance of 650–985ft (200–300m), but the cosmonauts would necessarily have to take over control for the final approach and docking. It was able to provide comprehensive control functions using complex algorithms and very large quantities of information processed on board as an integral component of the system of orientation and motion control (SOUD, in Russian). Designed by the Moscow-based NII-648 institute under their chief designer Armen S. Mnatsakanyan, it was thoroughly tested in an anechoic chamber during 1965 and 1966, where a great many potential defects were eliminated.

The ability to rendezvous and dock was crucial to operations in space and the effectiveness of the Russian system, developed for Soyuz, is a testimony to the high level of research and development in radio and electronic systems. In the early 1970s the Soyuz spacecraft was able to demonstrate an efficient means of sending crew and small quantities of cargo to the Salyut space stations, an investment that would pay dividends as the system was modified for more ambitious operations. The Americans too capitalised on their rendezvous operations but never developed an automated system. So it was with manual rendezvous technology that their Apollo and Skylab space station flights were conducted in the first half of the 1970s.

LEFT The docking probe on TMA-08M shows clearly the interface positions for electrical and mechanical connections as it backs away from the Poisk module at the ISS on 10 September 2013. *(NASA)*

By this time the Russians were demonstrating orbital rendezvous and docking as a fundamental step in their own space programme, now looking in a very different direction to the Moon. In selecting ever bigger and more capable Earth-orbiting space stations, they had the perfect vehicle in which to service and supply the facilities, with goods and crew members, in the decades to come. It was fitting indeed that in 1975 the Americans and the Russians came together for a docking in orbit of their Apollo and Soyuz spacecraft, a prelude to combined operations in space through Mir and the International Space Station.

In 1986 Soyuz TM spacecraft began using the new Kurs-A system for fully automated rendezvous and docking, enabling unmanned ferry vehicles in the Progress series to carry equipment to the Mir complex and later to the International Space Station. Kurs works in principle much like the Igla system on earlier spacecraft, with multiple antennae that work cooperatively with Kurs-P sensors on the docking ports at the Russian side of the International Space Station to effect a fully automated docking.

The system was efficient for Mir station operations in that, unlike Igla, it did not require line-of-sight orientation, thus saving propellant on the station, which now did not have to align itself with the approaching Soyuz to establish a link between the two vehicles. Moreover, Kurs-A could acquire the target at a range of 125 miles (200km) and fly around the station in close proximity to dock automatically at the assigned port. For a time, both Kurs and Igla systems were installed on Mir, as some modules sent to that complex had been intended for an earlier station which had been installed with Igla equipment.

In 2012 a developed version began using Kurs-NA antenna systems, which use less power and converge the work of the five antennae from Kurs-A into one. Because they protruded beyond the forward docking collar, Kurs-A antennae had to be extended after launch and retracted prior to docking.

ABOVE Oleg Kotov emerges from the Soyuz orbital module after TMA-10M docked with the Poisk module on 26 September 2013. *(NASA)*

LEFT A docking from a different perspective, as TMA-07M approaches the International Space Station on 21 December 2012 displaying detail on the docking unit. Note the Kurs rendezvous radar and docking antennae. *(NASA)*

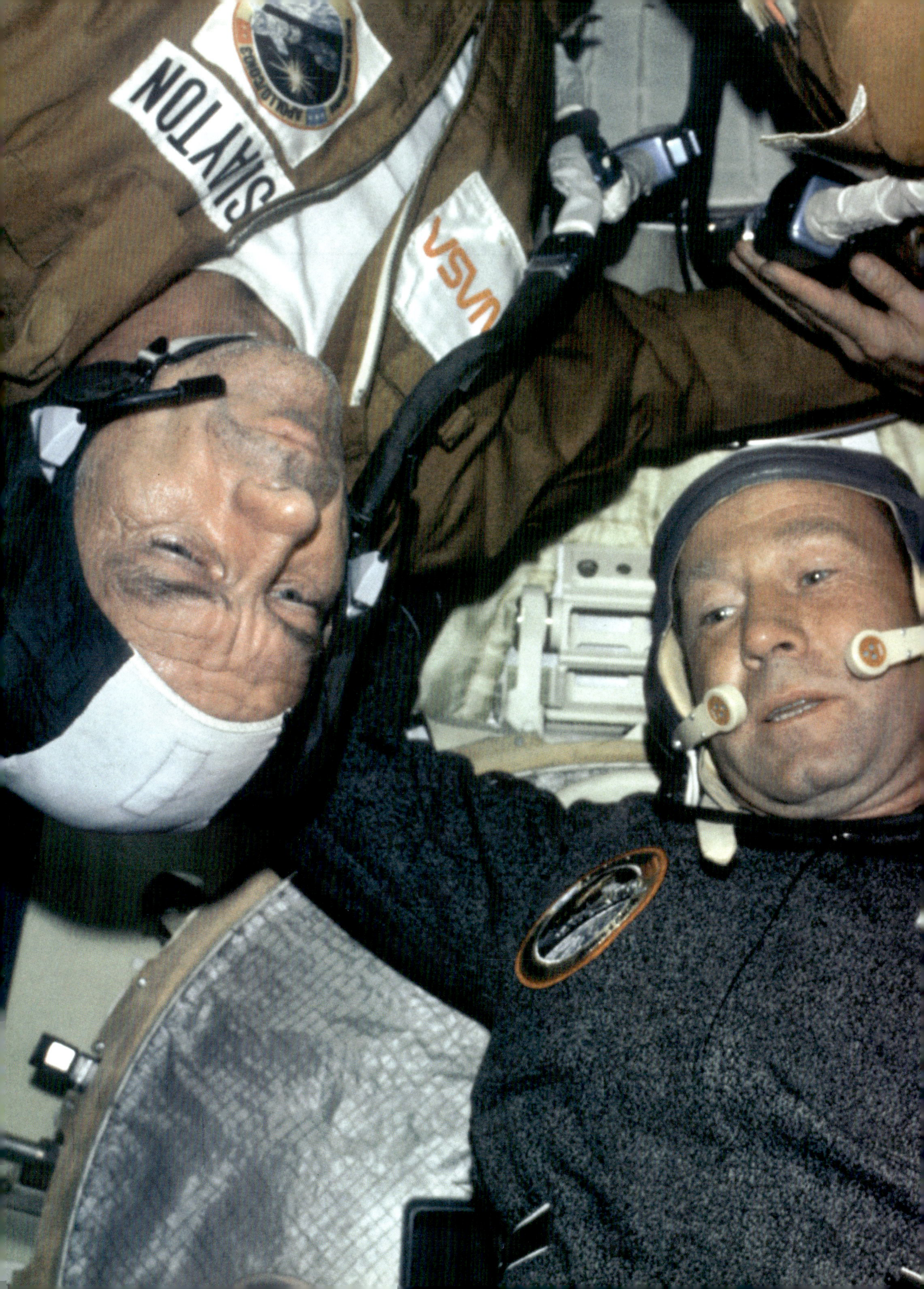
SLAYTON
NASA

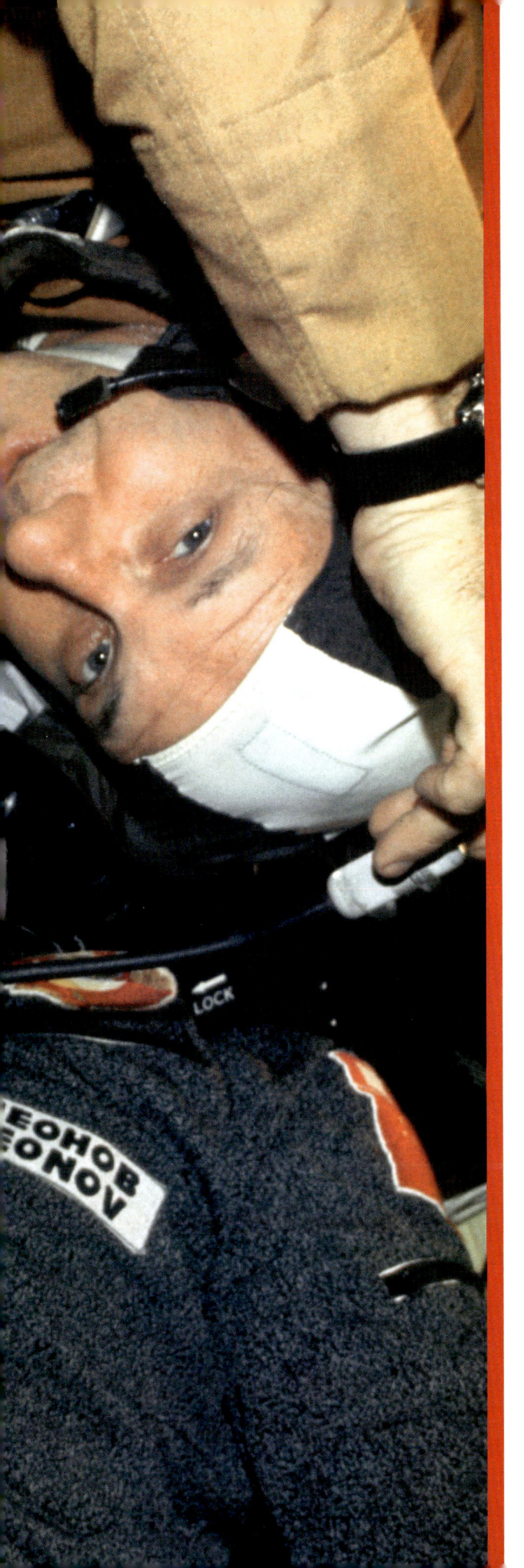

Chapter Five

Soyuz 7K-T ferry flights 1973–81

One of the most important changes to Soyuz was in the docking system that would allow cosmonauts to move from the orbital module to a space station without going outside on an EVA. The early 7K-OK systems had been adopted as a demonstration model and had not matured beyond bulky equipment fixed in place that could not be removed or redesigned to allow free passage through the forward hatch.

OPPOSITE From left to right, Mercury astronaut 'Deke' Slayton, Alexei Leonov and Tom Stafford, weightless in space and loving every minute! *(NASA)*

129

SOYUZ 7K-T FERRY FLIGHTS 1973–81

ABOVE By the time Russia had redirected the Soyuz programme to one that would support space stations in Earth orbit, the design of the spacecraft had stabilised into the configuration still recognisable today. *(Nick Stevens)*

The new 7K-T Soyuz would receive an active docking cone with a probe that would snag a conical receptacle on the station's docking port. Designed by the Azov Machine Building Plant, it was designed and tested in record time and provided a circular transfer hatch with a diameter of 2.6ft (0.8m). The life support systems were reduced in size and weight accordingly, although three crew members in shirtsleeves would fly the Soyuz, as they had with all Russian manned flights since Voskhod 2 in October 1964.

Major improvements too were made to the Igla rendezvous system, the equipment now being relocated from the toroidal base of the equipment section to the internal structure of the orbital module. The entire redesign was based on the use of existing technology, adapted or extended in capability for the new version. Because of this it was decided not to fly automated test vehicles but to move immediately to manned flights for orbital shakedown missions.

Despite weight-saving measures, the 7K-T was heavier than the 7K-OK, weighing in at an average 14,780lb (6,700kg), and had a length of 22.9ft (6.98m). It was no purpose-built ferry and clearly had a legacy of former roles for which it had never been flown. As a result, the download capability was very low, the descent module only carrying a maximum 44lb (20kg) of scientific equipment from any station to which it had been docked. Within three years engineers would rectify that limitation with the development of the 7K-TG, or Progress, unmanned cargo tanker.

The first flight of the 7K-T took place on 22 April 1971 with the launch of Soyuz 10, carrying Shatalov, Yeliseyev and Nikolai Rukavishnikov. The last had trained for a circumlunar mission. They were to rendezvous and dock with the first DOS space station, named Salyut 1 (which had been launched two days before by a Proton rocket), enter the station and remain for up to four weeks. There was again considerable pressure to get the first laboratory into space and this showed in the difficulties encountered with equipment.

Salyut

Russia's first space station, Salyut 1 comprised three cylindrical sections attached to each other to form a series of open sections with the maximum diameter of 13.6ft (4.15m) and a length of 51.8ft (15.8m) providing a habitable volume of 3,178ft^3 (90m^3). From the back to the front where the docking port was located, each section reduced in diameter, the centre section being 9.5ft (2.9m) across attached to a narrow forward section. In orbit Salyut 1 had a mass of 40,680lb (18,450kg) and a span of 33ft (10m) across the solar arrays. Attached either side of the station, one pair was located on the aft section and another pair on the forward compartment. Mishin

RIGHT The Luna series of unmanned Moon explorers would continue the work Russian scientists had hoped cosmonauts would conduct, a new role for robotic spacecraft epitomised here by Luna 16, the first of its type to return samples to Earth. *(David Baker)*

had wanted to call it Zarya (Dawn) but there had been a Chinese satellite with that name, so Salyut (Salute) was chosen instead. Each station initially had a life of 90 days but by the third station that had increased to 180 days.

Despite several technical issues right after launch, Soyuz 10 conducted a normal rendezvous with Salyut 1 and snagged the probe in the drogue, representing a 'soft dock' prior to driving the two circular docking rings together for a 'hard dock'. But that never came, despite Shatalov pulsing the thrusters to get a firm lock. Only then could they connect electrically to Salyut and prepare to open the hatches. Deciding to unlatch from a soft dock and try again, the crew discovered that they could not release Soyuz from the coupling. The only solution was either to pressurise the orbital module and go inside to manually disengage the docking system, or separate from the OM and leave it hanging on the end of Salyut 1's only docking port, in which case it would render the station unusable.

After more than five hours, Shatalov managed to get Soyuz 10 to undock, and with only one day's oxygen/nitrogen remaining in the environmental system, the only recourse was to come home. After an unspectacular re-entry, Soyuz 10 landed at night, another first for the Soviet space programme, at an elapsed time of 47hr 45min 54sec. Analysis revealed damage to the docking equipment caused by excessive accelerations at the initial docking attempt, some minor modifications, plus a change in operating procedure, appeared to solve the problem.

Soyuz 11 was to launch to Salyut 1 with Leonov, Kubasov and Kolodin and be followed by Soyuz 12 carrying Dobrovolsky, Volkov and Patseyev, but a medical check shortly before the first launch revealed Kubasov to have a suspected lesion on his lung. The result was that the crew for the second flight got moved up to the first. Soyuz 11 was launched on 6 June 1971 and, after a terminal rendezvous impeccably conducted by the Igla system, docked with Salyut 1 just 2hr 50min after lift-off. Patseyev was the first to enter the station, turning on some air ventilation equipment that had malfunctioned after launch.

The crew conducted their duties on a

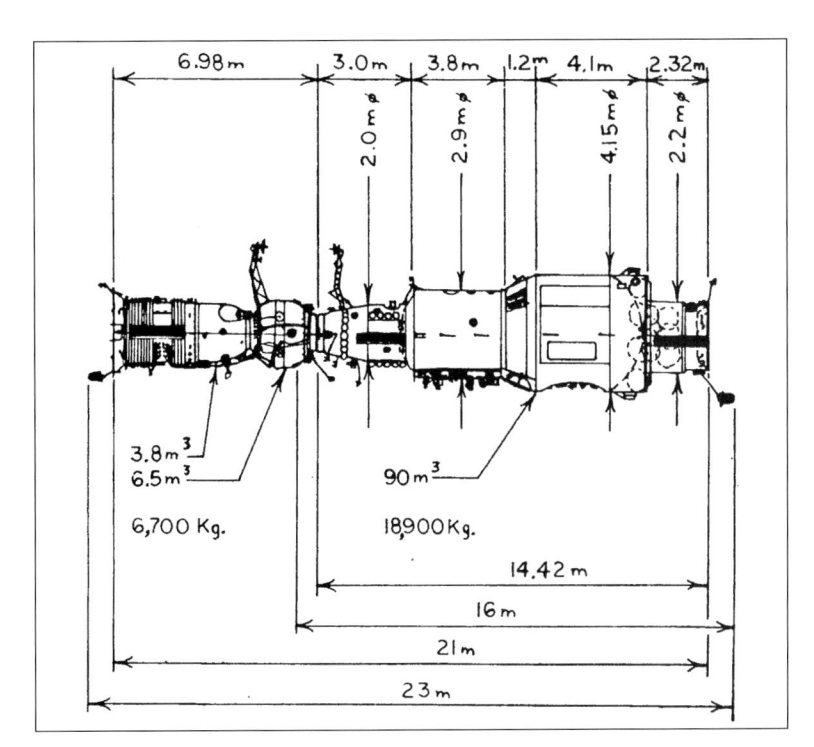

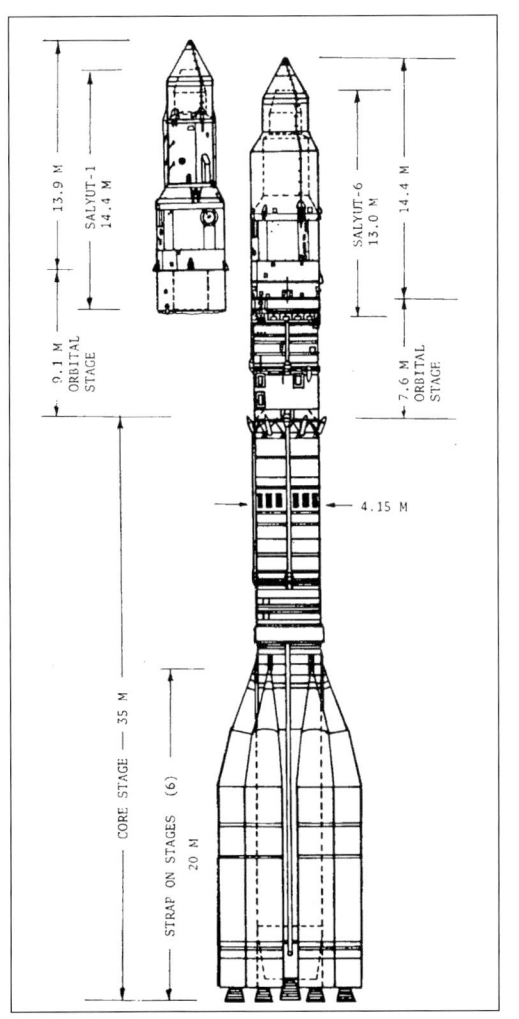

ABOVE Salyut 1 represented a major step forward for Russian space engineering and would provide the first habitat capable of receiving visitors in Soyuz spacecraft. As a station it was two years ahead of NASA's Skylab station. *(Charles P. Vick)*

LEFT The Proton launch vehicle would be employed for lifting to orbit all the Salyut civil and military space stations and their associated heavy modules. *(Charles P. Vick)*

RIGHT The Soyuz 7K-T ferry ship with capacity for a crew of three and modified solar cell configuration, displaying its tower-mounted rendezvous radar antenna. *(David Baker)*

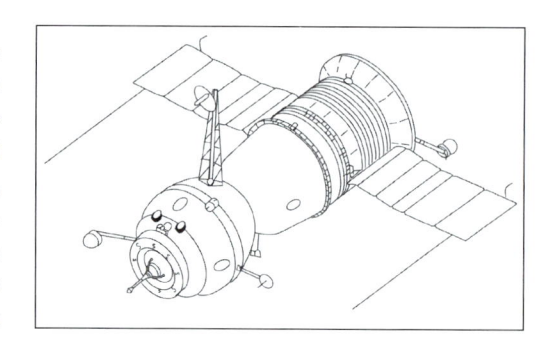

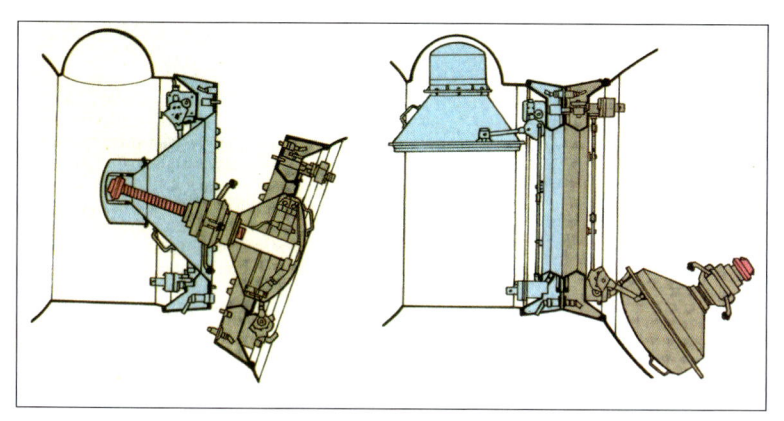

ABOVE The new docking probe and drogue, simplified and much lighter than the system used on early Soyuz rendezvous and docking flights, permitted the passage of cosmonauts between one facility and another without going outside. *(David Baker)*

rotational basis, with two always awake as the third slept, dividing the day into eight hours of work, two hours for meals, two hours for exercises, two hours of personal time and ten hours of sleep. With scores of science experiments to carry out and observations to make the flight was as much an endurance test as it was a scientific expedition, and psychological differences began to show. As the mission progressed tensions grew between the crew, a personality clash between Dobrovolsky and Volkov threatening to imperil order aboard the station. With tensions calmed and the mission running its course, the crew returned to their spacecraft and Soyuz undocked after 23 days 20 hours 40 minutes hooked up to the world's first orbital research laboratory.

After a total flight duration of 23 days 18 hours 21 minutes 43 seconds, the Soyuz 11 Descent Module landed as planned approximately 202km (125 miles) from the town of Dzhezkazgan in Kazakhstan. Tragically, all three men were dead inside their spacecraft. After an exhaustive review of what could have gone wrong, it was concluded that the sudden shock of separating the three modules after the de-orbit burn had caused a vent valve to open which normally would only have activated when the capsule was far down into the atmosphere and air could get in to balance the pressure. It had completely evacuated the pressurised compartment in 112 seconds, by which time the crew had been dead for at least 63 seconds. The lack of pressure suits had given them no chance of survival.

The immediate response was to halt all future flights pending modifications to the spacecraft, and to mandate the use of pressure garments. They were to be equipped with Sokol K suits which would be insufficient for a spacewalk but would ensure survival in the event of total depressurisation of the cabin. For reasons of weight, this would reduce from three to two the number of crew that could be carried, and not for more than nine years would another three-man crew fly in a Soyuz, when the T-series began flight tests in November 1980.

Another major change to the 7K-T was the reduced lifetime of the ferry ship, since it was designed to fly to and from a facility already in space and would not spend a lot of time in independent flight. For this reason the solar

RIGHT The docked configuration of Salyut 1 and Soyuz 11, the first transfer of crew members to an orbiting research station, manned for almost 24 days in April 1971. *(Charles P. Vick)*

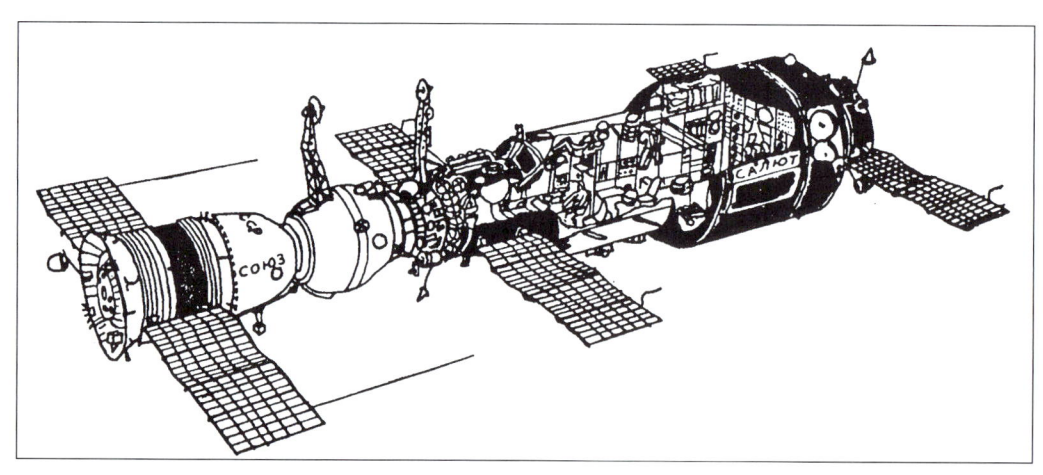

arrays were removed and electrical power provided by batteries. Once it arrived at the station, Soyuz would recharge those batteries from the station's electrical power supply. As designed, it had sufficient power for three days of independent flight and was rated for an orbital lifetime of 60 days docked to the station. In addition, two long whip antennae extended from opposing sides of the equipment module.

All these changes took weight away from the spacecraft but added greater mass from heavier items added, particularly the batteries and the new electrical distribution system necessitated by changing from solar cells. The net result was that the T-series weighed on average 15,000lb (6,800kg), 220lb (100kg) more than the earlier T-series used for Soyuz 10 and 11. These second-generation T-series Soyuz were the first to show development potential and each spacecraft would be started on a production-line basis but finished specifically with individual itemised pieces of equipment for a particular mission.

Second-generation K-T series flights

Salyut 1 had decayed down into the atmosphere on 11 October 1971 and the plan was to resume flights with the DOS stations in 1972, 1973 and 1974, with each receiving three or four visits. The decision to focus manned space flight on Earth-orbiting space stations provided an opportunity for the military to integrate Almaz stations with the civilian DOS stations built by Mishin. Each of the Almaz military stations would have recoverable film capsules for return to Earth in Soyuz. In the meantime, spacecraft 7K-T-33 was launched as Cosmos 496 on 26 January 1972 – to evaluate the modifications made after the loss of the Soyuz 11 crew – in a perfect flight lasting six days.

The launch of DOS-2, the second station, took place on 29 July 1972, but a failure in the Proton brought it crashing to destruction before reaching space. Amid the publicity surrounding NASA's last two Moon landing missions that year, the Russians felt it would be too minimal a demonstration of their capabilities to fly Soyuz 12 as merely a solo flight in Earth orbit. Chelomei

ABOVE The crew of Soyuz 11, left to right: Viktor Patseyev, Georgi Dobrovolsky and Alexander Volkov. Dobrovolsky and Patseyev had a personality clash on board, alerting physiologists to the ever-present need to carefully match crew members for long-duration flights. *(TASS)*

LEFT A commemorative postage stamp issued following the deaths of the Soyuz 11 crew when returning from their mission aboard Salyut 1. *(Novosti)*

had been persistently working on the Almaz military space station and it was this that saved the day when it was delivered for launch in early 1973 along with Mishin's next DOS station.

In an unexpected and somewhat controversial agreement, Mishin had agreed that his DOS stations would cease production after the fourth launch and that future space station activity would migrate exclusively to Chelomei's Almaz station, superficially similar to DOS. Initially, variants of the Soyuz would be used on these stations as ferry vehicles, but ultimately Chelomei's new manned spacecraft, TKS, would replace Soyuz altogether. It was proposed to Sergei Afanaseyev, the head of General Machine Building, and approved.

The reason Mishin agreed to this was so that he could focus on a new, expanded Moon programme, the L3M, which envisaged long expeditionary excursions to lunar surface bases. Mishin also wanted to use an upgraded version

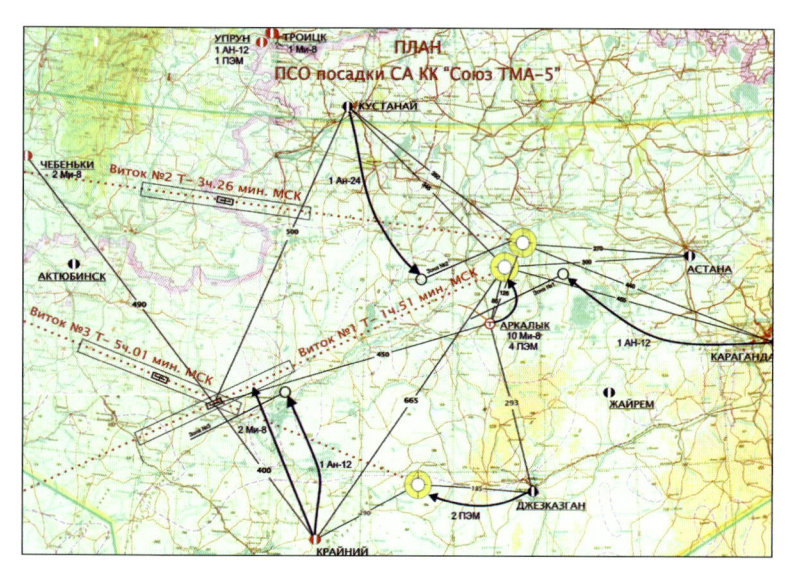

ABOVE The meticulous planning for recovery operations and the painstaking attention to the detail of entry path and descent trajectory is displayed on this map of the recovery area for Soyuz TMA-5, a common requirement of all flights. *(David Baker)*

of the N1 to support a series of manned missions to Mars, using elements of his existing Soyuz programme to support Earth-orbit assembly of massive ships. Needless to say, none of that had any fundamental basis in reality, although the political leadership in the Soviet Union had championed the improbable aspiration.

The first Almaz station, 101-1, was delivered to Baikonur in January 1973 and launched as Salyut 2 on 3 April on a mission during which it was to have received at least three visits. It was listed as a 'Salyut' to avoid suspicion that it was a different station to the civilian Salyut and therefore might have military functions. Almost as soon as it reached orbit things began to go wrong and the station was damaged by a series of technical problems, malfunctioning rocket motors causing physical damage to the workshop hull and the solar array wings being ripped off by an unknown event on the station. Unable to support manned visits it re-entered the atmosphere on 28 May.

With NASA about to launch its Skylab space station, Russian engineers were obliged to put on a sprint and get the next Salyut, one of Mishin's civilian stations, into space. Launched on 11 May 1973 it was placed in orbit but suffered from a faulty attitude control system, and when the orbital manoeuvring thrusters fired all the propellant to depletion it too was rendered unusable and re-entered the atmosphere a week after launch, remaining for many years in anonymity as Cosmos 557.

The technical failures to Salyut 2 (Almaz) and Cosmos 557 (Salyut) were an embarrassment at a time when final plans were being drawn up for a joint docking flight with NASA's Apollo spacecraft. To qualify the new T-series Soyuz variant, it was decided to fly a solo mission to test the new configuration, devoid of solar cells and with several new and untried components. That took place on 15 June 1973 when a Soyuz spacecraft designated Cosmos 573 performed a near flawless flight and landed just over two days later having qualified the new design.

This was followed on 27 September by Soyuz 12, which lifted into space carrying Vasily Lazarev and Oleg Makarov. It was the first time cosmonauts had been in space since the loss of the Soyuz 11 crew two years and three months before. The qualification flight went according to plan and lasted 47hr 55min 35sec, the first of two Soyuz missions put together to maintain an operational flow and to extract the most out of the successful Soyuz programme while the Salyut and Almaz space stations were working toward a restoration of flight operations.

With a new qualification for long-duration stays docked to a space station, the T-series brought new operational procedures, and the need to qualify these ferry vehicles for 60 days in space meant a Soyuz was launched on 30 November 1973, disguised as Cosmos 613. It simulated a rendezvous operation with a phantom target and shut down as if it had docked and was being configured in passive mode. The spacecraft performed well and returned to Earth on 29 January 1974, by which date another Soyuz had been launched with a crew.

On 18 December 1973, Soyuz 13 was launched with Petr Klimuk and Valentin Lebedev on board for an almost perfect flight lasting 7 days 20 hours 55 minutes 35 seconds, in which a very wide range of scientific activities were carried out in the orbital module. Designed to emphasise the new approach to manned space flight goals, the mission highlighted a diverse range of activities cosmonauts could do in space with potential benefits to people on Earth, mirroring a similar shift in American messages about the value of expensive and highly complex manned flight.

Research in orbital space stations was now becoming the focus on both sides of the ideological divide, as NASA worked to put the first reusable Space Shuttle in orbit within the next few years and aimed for a permanently manned space station. Operations with the Skylab space station were proceeding as planned and there was even an extension for the third and final flight from a nominal 56 days to a full 84 days in orbit, a mission that would be completed in early February 1974.

On 25 June 1974 an Almaz station, hailed as Salyut 3 to hide its identity, was launched by Proton rocket, followed eight days later by Soyuz 14 carrying cosmonauts Popovich and Artyukhin. The good fortune that had attended Soyuz flights since the tragic flight of Soyuz 11 was sustained for this mission, the first completely successful Russian space station operation, during which the crew spent two weeks aboard this military facility before returning to Earth.

Launched on 5 April 1975 with Vasily Lazarev and Oleg Makarov on board, what was later designated as Soyuz 18-1 resulted in the first high-altitude abort in any manned space programme. Devised as a 60-day visit to the Salyut 4 station, trouble began at an elapsed time of 4min 48.6sec when the second stage failed to separate cleanly from the core stage of the launch vehicle. Violently thrown free when the bolts finally wrenched loose, deviation from the upper stage's planned trajectory triggered ignition of the abort procedure at 4min 55sec.

Having already jettisoned the escape tower, the complete spacecraft separated from the offending stage as planned and conducted the standard abort procedure of igniting the main engine followed by separation of the three modules and a ballistic re-entry. Because the

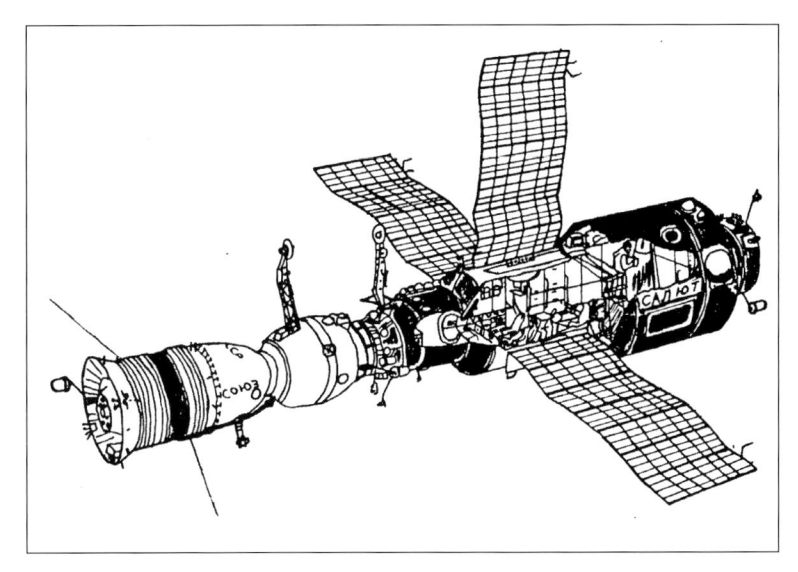

ABOVE Launched in December 1974, Salyut 4 (DOS-4) was host to Soyuz 17 and 18, the former depicted here docked to the forward transfer section. Note the third solar array wing. *(Charles P. Vick)*

BELOW Seen here with Soyuz 18, the general configuration of Salyut 4 was almost identical to Cosmos 557 (DOS-3), which failed due to a systems error. *(Charles P. Vick)*

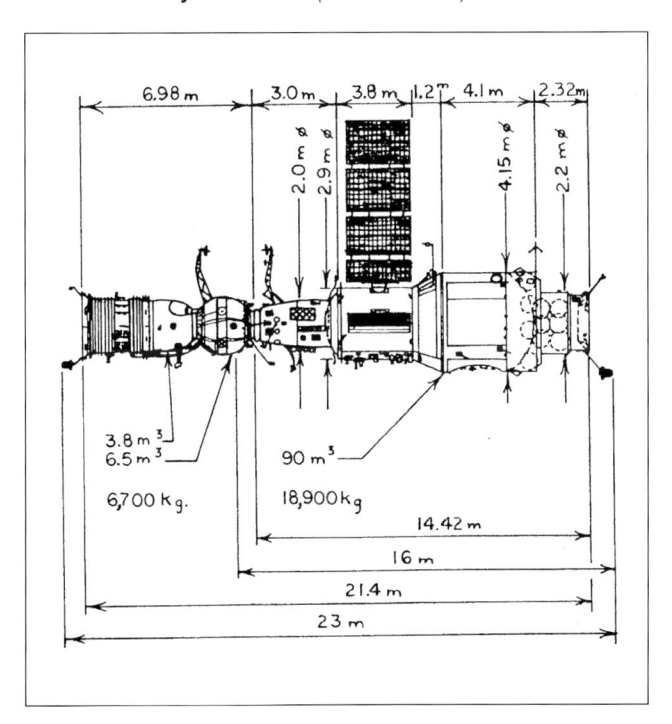

BELOW Following Salyut 2 (OPS-1) and 3 (OPS-2), Salyut 5 (OPS-3), launched in June 1976, was the last of the three military stations built as an Almaz facility but coded under the 'Salyut' name to hide its specific operational role. It utilised recovery capsules for returning film and other containers. *(Charles P. Vick)*

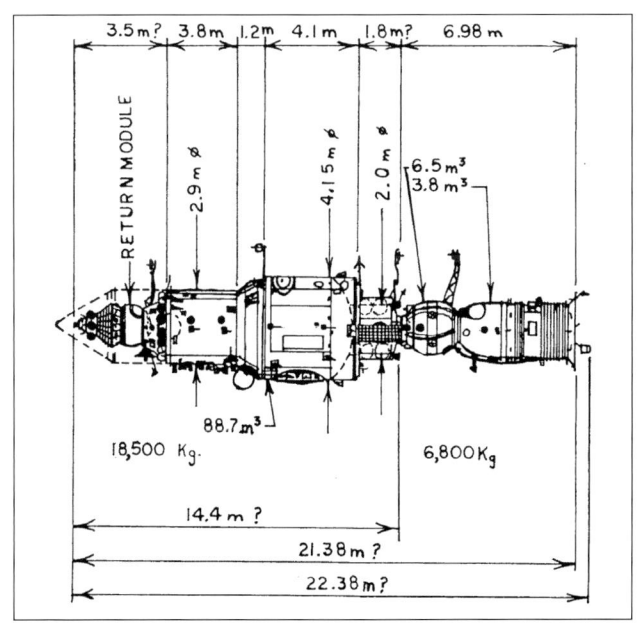

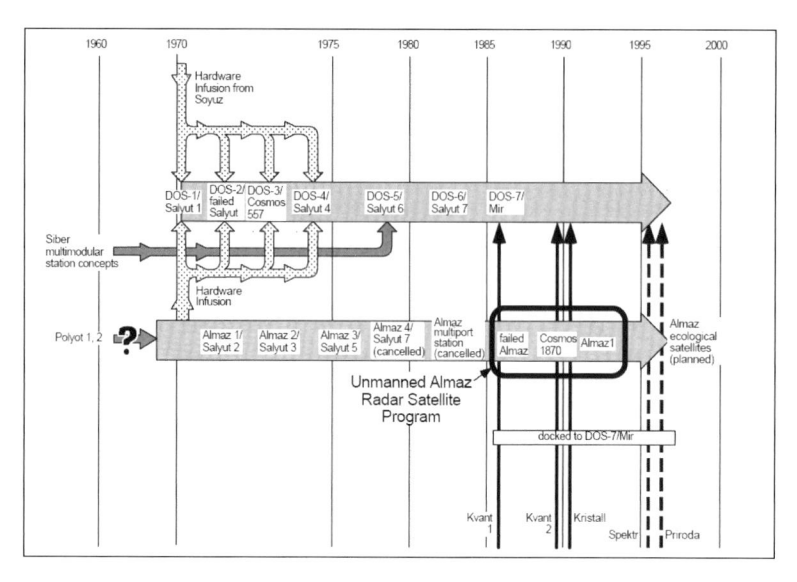

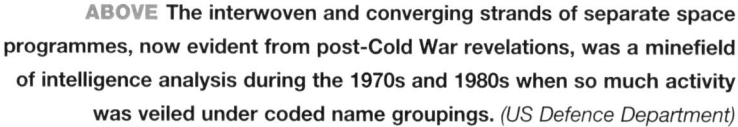

ABOVE The interwoven and converging strands of separate space programmes, now evident from post-Cold War revelations, was a minefield of intelligence analysis during the 1970s and 1980s when so much activity was veiled under coded name groupings. *(US Defence Department)*

ABOVE RIGHT Launched in September 1977, Salyut 6 with a Soyuz ferry vehicle attached represented the first really successful station of its class. Visited by 16 Soyuz vehicles, it supported five long-duration missions and eleven brief visits with short-duration crews. The longest single mission extended human space flight to 185 days. Salyut 6 was also the recipient of Chelomei's TKS vehicle. *(TASS)*

BELOW Salyut 6 had the ability to receive Progress unmanned cargo-tanker supply ships at the aft port, while receiving Soyuz ferry vehicles at the forward transfer compartment. *(Charles P. Vick)*

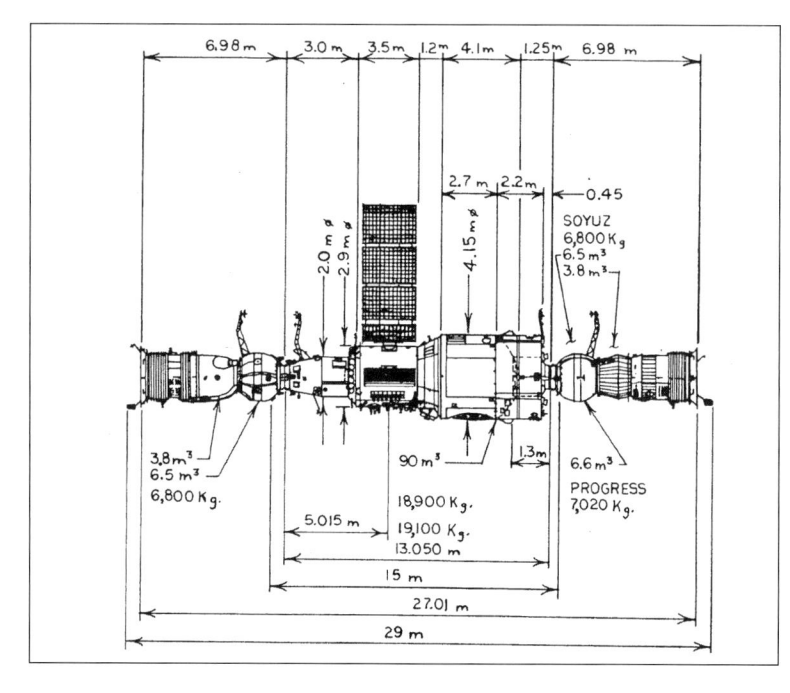

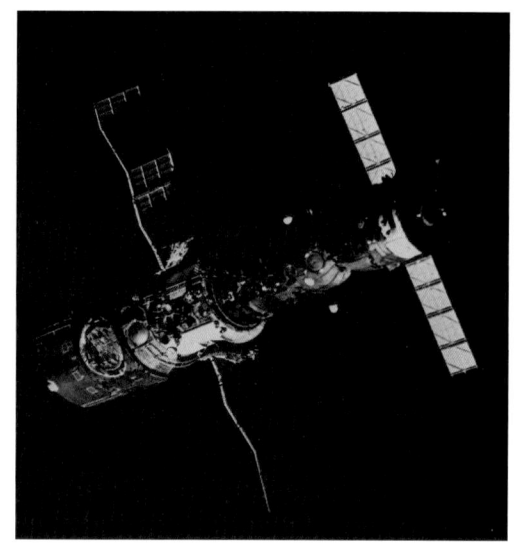

spacecraft was tilted down when the 'braking' manoeuvre was performed, instead of slowing the spacecraft it accelerated it, increasing the entry loads to a crushing 21.3g. Nevertheless, the parachute system worked and the crew landed just 515 miles (829 km) north of the border with China.

In a break with the tradition of secrecy surrounding failures the Russians gave full disclosure to the Americans, who were even then in the final months of preparing for the joint ASTP flight docking their Apollo to a Soyuz. But it was a failure to an older launch vehicle and in several ways validated the robust and reliable abort procedures and the survivability of the spacecraft under stresses far greater than any anticipated during test. A 'normal' abort would not have anticipated such severe forces acting on the structure.

In all, two visits were made to the Salyut 4 station, launched on 26 December 1974 and to which Soyuz 18-1 had been launched, first by Soyuz 17 in January 1975 and then by the crew of Soyuz 18 three months later. Intentionally brought down on 3 February 1977, it was succeeded by the military (Almaz) Salyut 5 launched on 22 June 1976. This was the last military Salyut launched, visited by Soyuz 21, 23 and 24 before it was de-orbited on 8 August 1978.

Launched on 29 September 1977, Salyut 6 was one of Russia's long-duration stations, remaining in space until de-orbited on 29 July 1982. In the intervening four years and ten months it received visits from 18 Soyuz spacecraft including four of the T series introduced in 1979, and the first 12 Progress cargo tankers adapted from the Soyuz spacecraft.

Toward the end of the K-T series, in response to NASA's announcement that it would fly teachers, journalists and non-professional astronauts from the scientific and commercial community, Russia began flying cosmonauts from friendly countries. Under the Intercosmos programme, candidates would be selected by a host country and put through a rigorous series of examinations before being selected for a space flight. The country would pay for the training and be assigned some level of scientific study or investigation of their choice.

The first of these Intercosmos Soyuz flights launched on 2 March 1978, with Soyuz 28 carrying the selected candidate from Czechoslovakia, Vladimir Remek, for a visit to the Salyut 6 space station. During the visit the resident crew delivered by Soyuz 26 passed the 84 days endurance record set by NASA in 1974 with the last Skylab visit. Remek's flight lasted 7 days 22 hours 16 minutes.

Between 1978 and 1988 Intercosmos flights continued with cosmonauts from Poland, East Germany, Bulgaria (twice), Hungary, Vietnam, Cuba, Mongolia, Romania, France (twice), India, Syria and Afghanistan. These flights included the K-T, T and TM series Soyuz transport ships. With the demise of the USSR in 1991, a new era of 'passenger' cosmonauts would open up in a very different political era.

The last of the basic K-T series was Soyuz 40, launched on 14 May 1981 with L. Popov and D. Prunariu on board, returning on 14 May 1981 following a visit to the Salyut 6 station.

ABOVE The spacecraft used for sending Hungarian Vladimir Remek to Salyut 6 on the first Intercosmos flight carrying non-Russian cosmonauts, Soyuz 28, in March 1978. *(David Baker)*

ABOVE The badge of Soyuz 30, which carried the Polish cosmonaut Miroslaw Hermaszewski into space to visit Salyut 6 in June 1978. *(David Baker)*

BELOW On display in a museum in Warsaw, Hermaszewski's Soyuz 30 descent module displays the parachute container housing, side window and forward port for access to Salyut 6. *(David Baker)*

BELOW The Soyuz 30 base section with aft heat shield jettisoned, revealing the four retro-rocket housings used to decelerate the spacecraft metres before touchdown. *(David Baker)*

BELOW The special medal struck to commemorate the flight of Mongolian cosmonaut Zhugderdemidiyan Gurragcha in March 1981. Launched in Soyuz 39, he was to spend seven days aboard Salyut 6. *(David Baker)*

Feature C

A handshake in space

ABOVE The mission badge for the Apollo–Soyuz Test Project, depicting the docked spacecraft with the crew names written in their native script, was a political statement commemorating a new period of co-existence rather than confrontation, endorsed by arms control agreements through the Strategic Arms Limitation Talks. The badge was designed and painted by Bert Winthrop of Rockwell International's Space Division. *(NASA)*

One of the most outstanding achievements of the space race was to promote new technologies that would be adapted as tools of international cooperation between states with ideological differences, in no small measure going some way to relieving tensions during the crisis years of the Cold War. Some have seen the Cold War as a reckless rush to Armageddon, averted more by luck than judgement. Others prefer to see it as a time when great tensions were held in check by the sheer magnitude of the consequences, while seeing a development in technology that allowed wars and conflict on a global scale to be averted on Earth while those same ideological differences were played out in space.

The origin of one of the more remarkable, yet frequently dismissed, aspects of the space race was the joint docking flight between a Soyuz spacecraft and an Apollo command and service module (CSM) in a project known as the Apollo–Soyuz Test Project (ASTP). Its roots lay in a series of cooperative agreements between the United States and the Soviet Union from the earliest months of the space age. Initiated by the agreement between many countries around the world to designate an International Geophysical Year of scientific study of the Earth and its environment, it helped forge links that would never be severed.

Sputnik and Explorer were products of the call upon countries to put up their technology, resources and scientific bodies to study the planet, resuming an effort initiated with the International Polar Year of 1882–83 and repeated in 1932–33. It was an attempt by science to heal the wounds of war and unite behind a common study of the Earth using a new range of technical tools including ships,

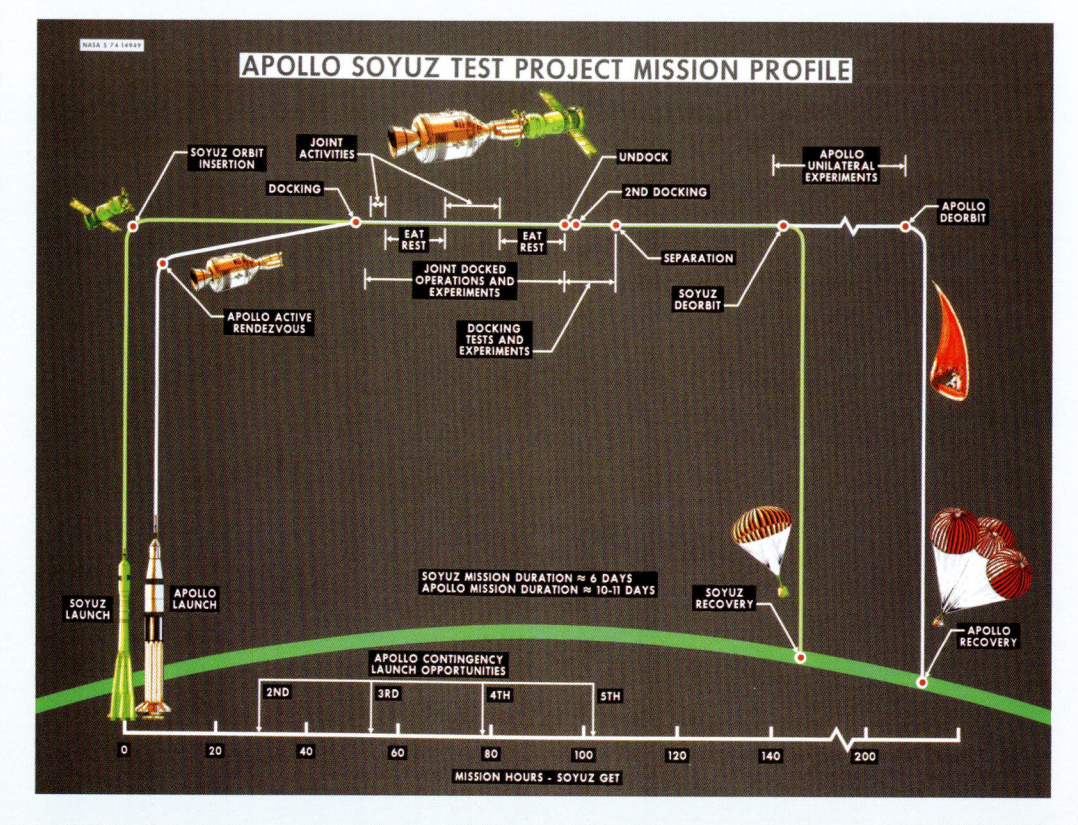

RIGHT A simplified artist's depiction of the planned link-up much publicised at the time and a representation of the new sense of cooperation. *(NASA)*

aircraft, rockets probing to the outer edge of the atmosphere and Earth-orbiting satellites.

With the growing desire to unite in studying near-Earth space and for exploring farther into the cosmos, several separate organisations got together to form the International Astronautical Foundation (IAF). Most of these were bodies formed in different countries to represent the scientific study of space and the development of rocketry, now seen as a practical possibility after the ballistic missiles of World War Two. One of those organisations was the British Interplanetary Society, formed in 1933 and still in existence today representing the space-faring interests of scientists, engineers, students and the general public in the UK and abroad.

Formed in 1951, the IAF became synonymous with the international scientific study of space and the planets, and it was through annual gatherings from many countries that dialogue began across the political and ideological divides. When Russia launched the space age and NASA was formed in 1958, government-level representation at annual congresses helped build bridges and secure the groundwork for agreements that would bind parties to principles upon which the exploration of space would proceed.

It was an opportunity for Russians and Americans to discuss areas of mutual interest and to secure working dialogue between parties. In America, the Office of International Programs was set up in 1958 as a branch of NASA responsible for these activities – the very Space Act which emerged from legislation setting up NASA required the agency to seek out ways of working with other countries on the peaceful use of outer space.

One of the first organs to emerge from this was the International Committee for Space Research (COSPAR) and much debate ensued in the United Nations about the levels of activity in space conducted by respective countries. The first several years were a time of competition and little cooperation, although by the early 1960s significant progress was made in sharing meteorological data from the ground and from satellites. But there were problems resulting from the very different way the two countries controlled their respective space activities.

America's civilian NASA programme was open to inspection, and separate from the government's military activities in space with satellites for spying, surveillance, snooping on radio signals and for communication, which was rapidly becoming an established part of US Air Force and CIA activities. NASA could afford to be magnanimous with offers of cooperation in the very separate sectors of science activity and human space flight.

The Russians controlled space activities through their defence infrastructure, not so much because they saw a tightly focused military purpose for all space activity but simply because they did not set up an equivalent organisation to NASA. Rockets were tied to military missiles – even Korolev's N1 had to be sold on the basis of its dual value as an asset for the military as well as for the propaganda value its missions would accomplish – and the manufacturing plants and launch sites all had dual functions involving national security.

Yet for all the apparent distance separating the two sides there were areas of common interest, and the IAF continued to play a big role in fostering bonds between scientists from East and West. It was through those efforts that the possibility of joint endeavours began to emerge. But set against a background of fierce Cold War interplay, it was remarkable that any progress was made at all. After Gagarin, everything changed.

Feeling a sense of technological superiority,

ABOVE A talented artist in his own right, Russia's Soyuz commander Alexei Leonov painted his own cartoon depiction of the Americans about to lasso the Soyuz spacecraft!
(Alexei Leonov)

RIGHT This configuration layout of the Apollo and Soyuz spacecraft together with the docking module manufactured by Rockwell International displays the size and weight of the Russian spacecraft. (Rockwell International)

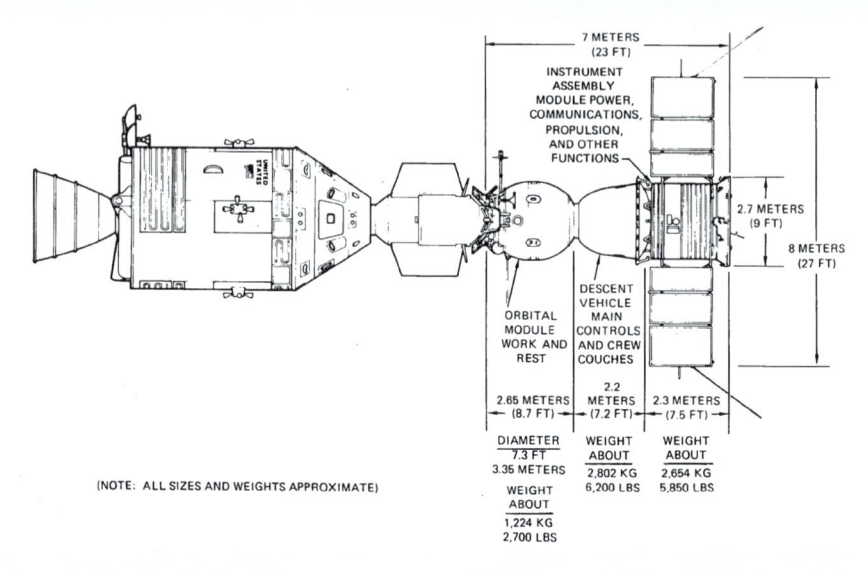

BELOW The APAS-75 androgynous docking equipment designed for the docking module and the Soyuz spacecraft was directly applicable only to the ASTP missions and related hardware. (Rockwell International)

the Russians could afford to be generous with offers of cooperation. Khrushchev was fascinated with the new US President, the charismatic John F. Kennedy, and talks began on a more serious vein when the Deputy Administrator of NASA, Hugh L. Dryden, met with Academician Anatoli Blagonravov, the Soviet representative on the UN Committee on the Peaceful Uses of Outer Space. From these discussions came an exchange in which the Kennedy administration expressed back-door interest in a cooperative Moon landing programme.

Horrified at the prospect of such an agreement, despite the fact that Lyndon Johnson reaffirmed the possibility of such a deal after Kennedy's assassination, Congress turned it down emphatically. And there the matter might have rested forever. But seeds had been sown. In 1964 the Soviet Union put great effort into its own Moon landing programme, with no result to show for it, while the Americans achieved their goal and put Armstrong and Aldrin on the surface of the Moon in July 1969. With a new political leadership in Washington and Moscow, fresh opportunities arose out of the final hurdles in a race that began more than ten years earlier.

Late in 1969, Soviet Academician Boris Petrov met with NASA Administrator Thomas O. Paine. The Russians saw the Americans as having achieved parity with them and there was a new feeling of equivalence from Moscow, that they understood that the Moon mission was something that NASA had to do to restore national pride after they, the Russians, had demonstrated superiority with Sputnik 1 and Yuri Gagarin. From a position of equality, as they saw it, the political balance was in equilibrium.

The loss of Soyuz 1 in April 1967, less than three months after three American astronauts lost their lives in a launch-pad fire, moved the then NASA Administrator Jim Webb to ask: 'Could the lives already lost have been saved if we had known each other's hopes, aspirations and plans? Or could they have been saved if full cooperation had been the order of the day?'

From the spring of 1969 an exchange of correspondence took place between president of the Soviet Academy of Sciences Mstislav Keldysh and Paine over areas where the two countries could work together in that most

FAR LEFT Design test shot of the androgynous APAS-75 unit with the upper component in active (extended) mode as it would be if it were about to lock on to the lower component. *(NASA)*

LEFT The engagement of the captive with the passive component of the APAS-75 unit displays the extended pistons that will retract to draw the two sides into a locked position. *(NASA)*

BELOW NASA engineers and their Russian equivalents discuss aspects of the engineering design for the androgynous docking unit in Building 13 at the Johnson Space Center, Houston. *(NASA)*

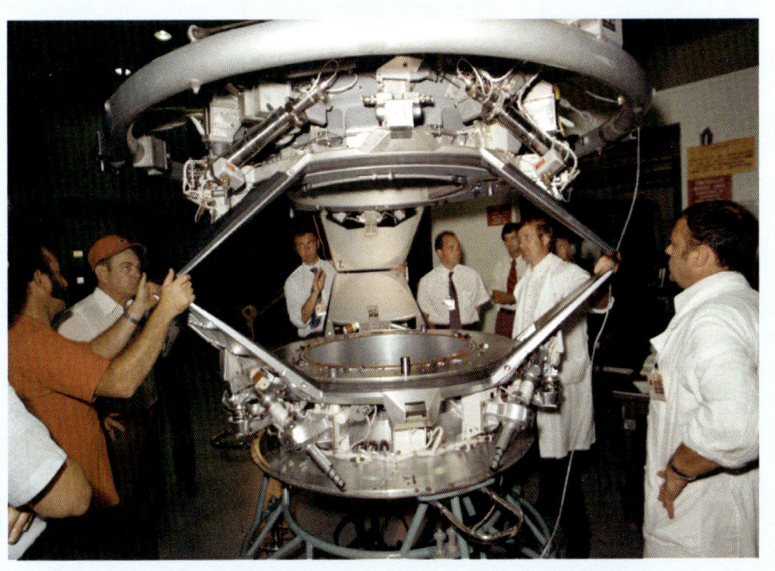

public and empowering demonstration of national pride, human space flight. Shortly after the near disaster to Apollo 13 in April 1970, the desire to work together seemed more poignant than ever. The White House was amenable to the idea and it fitted in with Richard Nixon's foreign policy. From a world made dangerous in the 1950s by naked confrontation, to strident competition in space throughout the 1960s, perhaps the 1970s would be the decade of mutual co-existence instead of 'mutually assured destruction', the catchphrase of deterrence.

Possible cooperation quickly focused on a compatible docking unit that could be used in life-threatening situations in space where one nation could come to the aid of another for humanitarian reasons. (Each country had developed unique docking apparatus and neither would work with the other country's vehicle.) And each spacecraft would require compatible communication, telemetry and tracking systems. As NASA engineers began to mull over possibilities, the difficulties mounted.

The existing Soyuz could carry only three suited cosmonauts: there would be insufficient room for a rescue crew plus three Apollo crew members escaping a stranded spacecraft. But Apollo could come to the aid of Soyuz, because there were already plans for the command module to carry five astronauts should that be needed for NASA's upcoming Skylab space station and sufficient room to take on a Soyuz crew as well. The extra two would be carried on couches below the standard seats.

The real work of trying to define a possible contingency rescue capability, and perhaps configure a demonstration flight to prove the possibilities, was conducted during the second half of 1970. In these several months, the Russians learned about NASA's plans for a Space Shuttle and a permanently manned space station. For their part the Russians were already demonstrating an expanding commitment to orbital laboratories for sustained occupancy. With NASA replacing one-flight spacecraft with a reusable transportation system, the logic of using space stations as a 'safe haven' for stranded crews made sense. It was this area on which attention was focused.

Much discussion prevailed around demonstration flights, and NASA's Skylab space station was mooted as a possible

141

SOYUZ 7K-T FERRY FLIGHTS 1973–81

RIGHT This schematic shows the interior of the docking module and displays the panel and container space for a variety of scientific experiments the US astronauts would conduct for a further three days after the joint-docking phase of the flight was over. *(Rockwell International)*

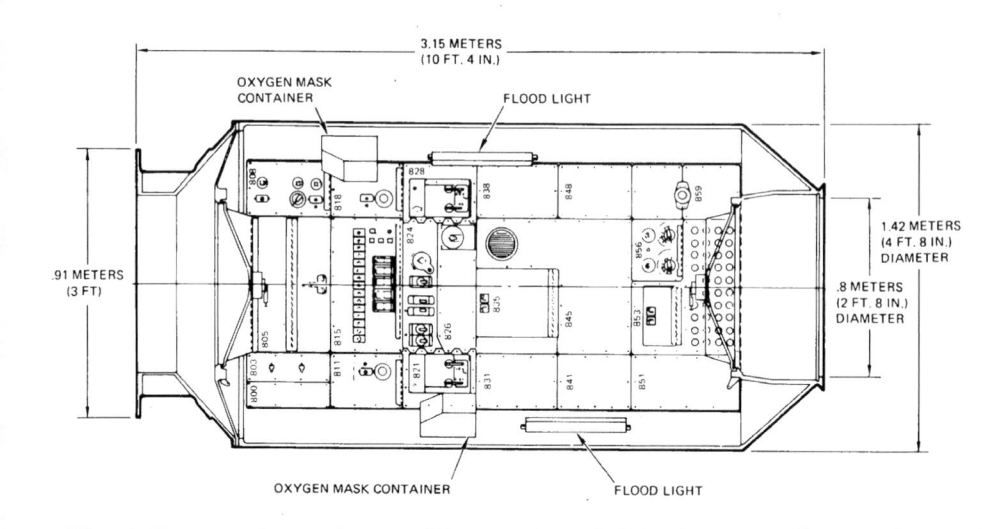

BELOW The hatch connecting the docking module with the Apollo command module had pressure equalisation gauges on each side, a vital determination in the use of the DM as a transitional compartment between the two spacecraft. Because of the different pressures and atmospheres in respective spacecraft, at no time was the DM open at both ends. *(Rockwell International)*

rendezvous for Russian and American crew members. Skylab, comprising a converted Saturn V rocket stage and a cluster of purpose-built modules, would be visited initially by an Apollo crew spending 28 days in orbit, followed by two further visits at intervals of 56 days each. But there were other difficulties too. American and Russian spacecraft used different environments: Apollo had a pure oxygen atmosphere at one-third sea-level pressure while Soyuz had a nitrogen/oxygen atmosphere similar to that at the surface of the Earth. Nobody thought the problems totally insoluble.

Full of optimism, a NASA team arrived in Moscow in October 1970 and was taken to Zvezdny Gorodok, Star City, a short car ride outside Moscow, where the cosmonaut training centre was located. Situated some distance off the main road, it was surrounded by trees within a small wooded area that effectively screened the place from prying eyes. At least two manned security barriers were encountered before the 1950s-style Soviet buildings suddenly appeared out of the trees. For the first time, senior NASA managers and engineers were getting a guided tour of Russia's highly secret cosmonaut complex.

The Russians showed off their training facilities – the centrifuge where candidates were tested for their tolerance to g-forces, simulators where missions were rehearsed and procedures tested, and mock-ups where the NASA people crawled inside the habitable modules that made up a Soyuz spacecraft. For their part the Russians learned how Apollo docking mechanisms worked, about the procedures used by the Americans for rendezvous and about the way their spacecraft was operated. It was done through interpreters, but the spark that came from enthusiastic engineers discussing their favourite subject was endemic.

In further discussions around desks, the Russians explained that the rudimentary docking apparatus that prevented cosmonauts from moving directly from one Soyuz to another while docked would be replaced by a more advanced system that would open a tunnel between the two spacecraft. No longer would cosmonauts have to do a spacewalk to get from one ship to another. Details of the Skylab space station were of particular interest to the Russians, who were engaged in a long development path towards their own, more ambitious, space station programme.

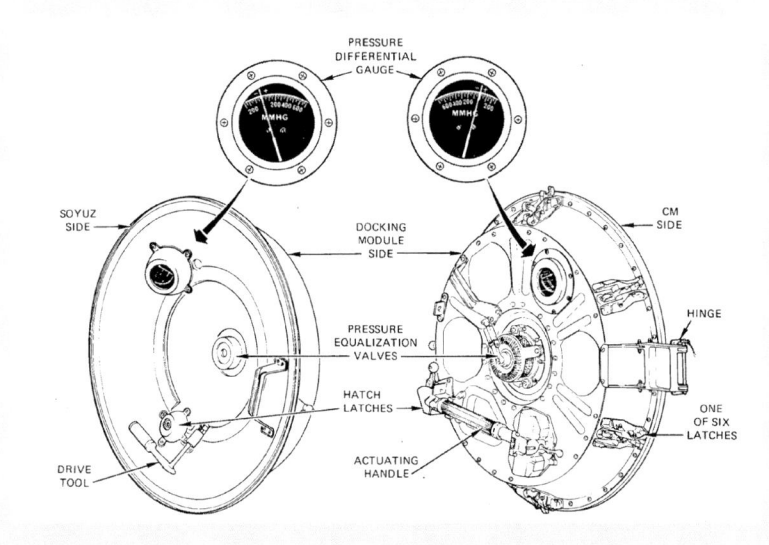

142
SOYUZ MANUAL

The Russians were fascinated to learn that by the end of that decade NASA hoped to have a permanent station with 12 people routinely served by the reusable Shuttle. All of which in fact was very public knowledge and much-discussed in the specialist press. It was against this background of rising hope for a more sustainable space programme, devoid of fast sprints for short stays on the Moon, that common ground was found to continue the discussions. Another visit was arranged for early 1971, this one including the NASA Deputy Administrator George M. Low.

At this next meeting, in January 1971, discussion focused around the prospect of a joint docking flight between Apollo and Soyuz, with extensive debate on how the incompatible environments of the two spacecraft could be accommodated – either by astronauts wearing lightweight suits or by a special module operating as an airlock, although not depressurising to a vacuum, but merely to avoid the Soyuz crew getting the bends by going straight to a pure oxygen environment. The talks got down to substantive detail, and a considerable amount of technical information was exchanged, most of which focused on docking systems and how to match the two environments. Quite suddenly, amid a series of discussions where the difficulties seemed to be compounding themselves, a solution emerged almost by itself.

A docking module (DM) attached to the front of the Apollo spacecraft could carry at its opposite end a docking system compatible with the Russian design. This would be the end to which Soyuz would be docked. With hatches at each end to allow access from, into or between each spacecraft, the DM itself could be the atmospheric airlock. The crew from the Apollo spacecraft would enter the DM, close the hatch and raise the pressure to 10lb/in² (518mm Hg), adding nitrogen to match the environment in Soyuz. Soyuz would then reduce its pressure to be compatible with the docking module and the hatch from the Russian side could be opened, allowing the crew to move across.

FAR LEFT Apollo ASTP commander Tom Stafford (standing) poses at the orbital module training mock-up at NASA's Johnson Space Center, while Alexei Leonov peers through the OM window. *(NASA)*

LEFT The descent module mock-up at the cosmonaut training centre outside Moscow, where Apollo astronauts familiarise themselves with the interior of the Soyuz spacecraft, provided a fast learning curve for Americans unfamiliar with Russian design concepts. *(NASA)*

LEFT Cosmonauts and astronauts accompany a US–Soviet team inspecting the Soyuz spacecraft that will be used in the joint docking flight. *(NASA)*

ABOVE Unique to this mission, Soyuz 19 is equipped with the androgynous docking system with which it will engage the paired unit of the same type on the Apollo docking module. *(NASA)*

LEFT A view of Soyuz 19 from Apollo CSM-111, which was launched 90 minutes after the Russian spacecraft, as rendezvous is achieved. *(NASA)*

BELOW In this artist's view, the APAS-75 system is depicted in passive mode ready for latching to the docking module. *(Rockwell International)*

To return, the crew would re-enter the DM, seal the hatch, lower the pressure to 5lb/in^2 (258mm Hg) while purging it of nitrogen, open the hatch and drift back into the Apollo command module and its pure oxygen environment. By lowering the pressure from the mixed-gas atmosphere of Soyuz, the crew would avoid the 'bends' so feared by scuba divers returning to the surface too quickly after a dive. All the necessary equipment for conducting the exercise could be contained in the specially outfitted DM, with different docking systems at each end compatible with the respective spacecraft. The DM would contain the oxygen and nitrogen tanks for pressurisation together with all the regulators and communications equipment necessary for sustained contact between spacecraft and with the ground.

With the general outline of an agreement to hand, both parties agreed to move ahead and a number of working groups were set up to deal with specific challenges posed by the plan. The next step was to get a formal Soviet delegation across to Houston to see how NASA did things and, no less important, to understand the culture that lay behind the Apollo programme. The Russians arrived in June 1971 and were quickly introduced by their hosts to how generous American hospitality can be! Barbecues, fast cars, pool parties and shopping malls were only some of the eye-watering experiences to which the Russian visitors were exposed.

The general mood of relaxed anticipation for a world-changing event came to a tragic end shortly after the Russians departed for Moscow when, on 29 June, the three-man crew of Soyuz 11 lost their lives returning to Earth after 23 days aboard the Salyut 1 space station. This came just three days after the third and very violent failure of the N1, an event that had been photographed from space by the Soyuz crew. Soviet morale sank and messages of condolence were sent from America to the Russian space community.

There were other reasons why America wanted a joint flight with the Russians, still not yet officially ratified by respective governments. From heady and lofty heights, the NASA budget was falling dramatically, and the Shuttle on which so much hope had been placed was running into budget problems – estimates on its development cost were simply too high. With the last two Moon expeditions planned for 1972, and Skylab in 1973–74, there would be a long gap before the Shuttle could begin operations, assuming it was approved. A joint flight would help plug that gap, maintain facilities at an operating level for a little longer and use existing hardware already built and paid for to exact that advantage.

But the joint flight with a Soyuz was not the only possibility on the agenda. Throughout the second half of 1971, discussion included thoughts about a docking mission to a Salyut space station and the development of a universal docking system that could be carried by the Shuttle. NASA had four Apollo spacecraft left over from the Moon landings, three of which had been cancelled, and raised the possibility of a more extensive series of flights. Within this period, the notion of a flight to Salyut, not a Soyuz, gathered credence and both sides were more enthusiastic for this type of flight. It would still involve a Soyuz but the prime objective would be the Salyut station.

In November 1971 the third joint meeting was held when 20 NASA representatives went to Moscow. The results of the working groups were aired and consideration was given to a schedule of events leading up to an actual flight. It was agreed that 1975 was the earliest this could take place. Discussion had already been held with North American Rockwell, builders of Apollo, about design and construction of the docking module, and the consensus was that this was the earliest possible date. Technical discussions were also held about an international docking system with some proposals from the Russians about possible configurations.

When the Americans returned home in the December, a recommendation was made to the White House that a formal declaration of the joint flight be a part of the planned summit between President Nixon and Premier Kosygin in May 1972. Henry Kissinger got involved and wanted clarification that there would be no technical obstacles to embarrass either side, should such a commitment be given. He was assured that there would not. But when they met in Moscow in April 1972, NASA was told by the Russians that Soyuz was to be the docking target after all, and not Salyut.

The April meeting had been a clandestine affair at the highest level, with very few people

BELOW The Apollo CSM-111 with the docking module attached to the docking ring on the command module using a standard Apollo-type probe and drogue. *(NASA)*

ABOVE An artist captures the scene of the seminal moment when astronauts and cosmonauts meet in space. *(NASA)*

being privy to the visit to Moscow made by George Low, but when Nixon met Kosygin at the planned summit the deal was done, signed on 24 May 1972. It was a five-year agreement involving several separate areas of cooperation, and this included the joint docking flight. Perhaps of greater note was the signing of the first Strategic Arms Limitation Talks agreement setting limits on the rate of growth within the nuclear defence arsenals of each side. But the joint ASTP flight was a strong sign of goodwill.

The agreement tied both sides to a flight in mid-1975. It was the first binding agreement between the United States and the Soviet Union, and linked two very different human space flight programmes in a common endeavour. It was made possible by Soyuz, which the Russians saw as having a protracted life stretching several decades into the future. For the Americans, it was to be the last flight of their Apollo spacecraft and the last time US astronauts would fly into space for several years. Almost five months earlier, in January 1972, President Nixon had formally given the go-ahead to the Shuttle, the agency having found a way to fit the design within the restricted development budget set by the government.

With a clear and very public commitment to cooperation, the hard work necessary to solve numerous technical problems and to develop a programme that would satisfy the requirements of both sides began in earnest. But for others it was a sign that America was going soft on its hard-line policy towards communism, and some elements of the press – including some specialist media – began a campaign to discredit the project. Questions were asked about the degree to which American technology was being 'given away' and how reliable the spacecraft was, critics citing the four deaths in just ten manned Soyuz flights.

NASA named the crew for the ASTP mission on 30 January 1973. The prime crew would consist of Thomas P. Stafford, veteran of one Apollo and two Gemini missions, as the commander, Vance D. Brand as command module pilot, and Donald K. 'Deke' Slayton as docking module pilot. Their backups were, respectively, Alan L. Bean, veteran of one Apollo mission and one Skylab flight (which was yet to take place), Ronald E. Evans, veteran of one Apollo mission, and Jack R. Lousma, also assigned to a Skylab flight. Slayton had been selected as one of the seven Mercury astronauts but had been removed from flight status when diagnosed with a medical condition that had since been rectified.

Russian crew assignments for the ASTP flight were announced in June 1973 and involved several cosmonauts, including Alexei Leonov and Valery Kubasov as the prime crew with Filipchenko and Rukavishnikov as their backup, plus paired crews Dzhanibekov/Andreyev and Romanenko/Ivanchenkov as candidates for a full dress rehearsal to qualify the modified Soyuz.

Over the next two years cosmonauts and astronauts learned each other's languages to varying degrees of competence, while technical groups visited each other's countries and worked on the finer details of how to make the mission a success. A great deal of scrutiny was applied by NASA to the design and engineering of the Soyuz spacecraft and much was learned by both sides. In many ways the Soyuz was still in a developmental stage, with many improvements being made for the resumption of flight after the loss of Soyuz 11. As recorded elsewhere in this book, manned missions resumed on 27 September 1973 with Soyuz 12 to flight qualify the redesigned vehicle.

ABOVE Leonov and Kubasov go through a checkout of the orbital module, as viewed by a NASA flight crew member. *(NASA)*

BELOW As seen in this frame from a 75mm movie, Tom Stafford and Alexei Leonov conduct their historic handshake across the open hatch. *(NASA)*

LEFT Tom Stafford, commander of the Apollo spacecraft, and Alexei Leonov, commander of Soyuz 19, converse with each other across the hatches separating the docking module from the orbital module. *(NASA)*

RIGHT In this depiction, one astronaut and one cosmonaut occupy the Soyuz orbital module and another pair is in the docking module, leaving the third NASA astronaut in the command module with the hatch sealed to maintain the different pressures and atmospheres. *(NASA)*

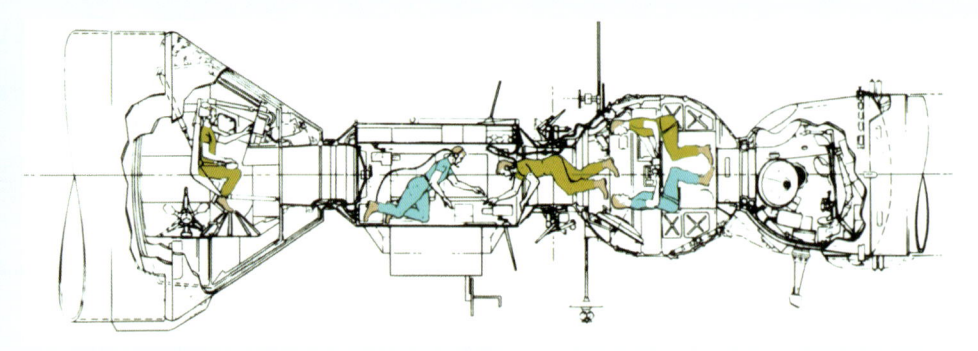

In direct support of the ASTP mission, and recovering from the Soyuz 11 disaster, the Russians planned to fly three test flights before the ASTP mission, one unmanned and two with crews on board. The first, the unmanned Soyuz, designated Cosmos 638, was launched on 3 April 1974 for a flight lasting almost ten days. This flight also introduced the improved 11A511U rocket, hereafter known as the Soyuz U launch vehicle. The modifications made specifically for the ASTP mission were evaluated in flight, including an adapted environmental control system capable of supporting four people inside Soyuz, as it would be called upon to do on the actual docking flight. This ship also had the extended solar array panels for the added power required, and was the first all-up test with the new universal docking equipment.

A minor problem during descent gave concern about the pressure equalisation valve and it was decided not to fly two manned precursor tests but to make the next flight a repeat of Cosmos 638. Coming soon after Soyuz 11, the engineers were sensitive to all pressure equalisation changes, even if it was a subtle variation and not a failure. Launched on 12 August 1974, Cosmos 672 was almost identical to its predecessor, complete with the new docking system, in a flight lasting 5 days 22 hours, duplicating the anticipated ASTP mission.

The full rehearsal with crew on board was launched on 2 December 1974, carrying Filipchenko and Rukavishnikov. On this flight, designated Soyuz 16, all ASTP events were rehearsed on an equivalent timeline and tests were made with changes to the pressure levels, adjusting the oxygen from 20% to 40% of the mixed-gas atmosphere when reducing the level to 540mm Hg. To evaluate the new APAS docking design, Soyuz 16 carried a copy of the US docking ring on the front of the orbital module, a copy of that which would be installed on the docking module. The crew practised extending and retracting it several times in flight before it was jettisoned prior to re-entry. The flight lasted just under six days, very close to the planned duration of the ASTP flight.

Considerable scrutiny surrounded the post-flight debriefings, to which the NASA engineers gave particular attention. The ASTP event was being televised and would bring worldwide attention and there were detractors both inside and outside the space programme, in both countries, waiting for their worst fears to be justified. In the event, the mission was passed as having achieved all its objectives, clearing the way for this heavily politicised flight. It is ironic that the President who would make the telephone call to the dual crew in space would be Gerald R. Ford, successor to the disgraced

RIGHT Mission Control in Houston monitors the joint activities aboard the docked vehicles in space, a time of reflection for some on the last Apollo flight of all. *(NASA)*

RIGHT President Gerald R. Ford talks with the ASTP crew as they orbit the Earth together during joint activities that lasted almost 48 hours. *(NASA)*

Richard Nixon, the man who had cleared the way for both ASTP and the Shuttle which would succeed Apollo.

Designated Soyuz 19 and weighing 14,970lb (6,790kg), the 7K-TM spacecraft was launched at 12:20pm Universal Time on 15 July 1975 with Leonov and Kubasov on board. The spacecraft was launched to an orbit of 137 miles (220km) by 115 miles (185km) before orbital manoeuvres placed it into its target orbit of 144 miles (231km) by 135 miles (218km), awaiting the launch of the Apollo spacecraft. Because Apollo had, for operational necessity, been designed to carry much a greater propellant load than the far smaller Soyuz, it would be the active vehicle, conducting all the rendezvous manoeuvres to catch up with Soyuz 19.

Weighing 32,558lb (14,768kg), the Apollo spacecraft (CSM-111) was launched at 7:50pm Universal Time carrying Stafford, Brand and Slayton on the last ride of the Saturn IB (SA-210), initially entering an orbit of 106 miles (171km) by 96 miles (154.5km), some 4,223 miles (6,796km) behind Soyuz. The first activity involved the Apollo spacecraft separating from the upper stage of the Saturn IB, turning around 180° and extracting the 4,436lb (2,012kg) docking module. Through a sequence of orbital manoeuvres conducted by Apollo, a rendezvous with Soyuz 19 was achieved and a docking accomplished 44hr 19min after it had lifted off from the Kennedy Space Center, 51hr 49min after the Soyuz launch.

The two vehicles remained docked for about 43hr 50min, during which several multiple crew exchanges took place using the docking module as an atmosphere transition compartment. The period was punctuated with handshakes, ceremonial exchanges of medals, halves launched separately to be united in space, telephone calls from the White House and shared meals. After undocking the two spacecraft went their separate ways, Soyuz 19 touching down on 21 July, 5 days 22 hours 31 minutes after launch, Apollo remaining in space and conducting various experiments before splashdown on 24 July, 9 days 1 hour 28 minutes after it had been launched.

For NASA it was the end of the era of one-mission spacecraft designs, and the last human space flight in which American astronauts would ride into space for almost six years. Their next would be the first flight of the Shuttle *Columbia*, launched on 12 April 1981, exactly 20 years to the day after the historic mission of Yuri Gagarin.

FAR LEFT Never far from a sketch pad, Alexei Leonov shows off his portrait of Tom Stafford. *(NASA)*

LEFT Bert Winthrop's painting of the Apollo–Soyuz crew members (from left to right: Donald K. Slayton, Vance Brand, Tom Stafford, Valery Kubasov and Alexei Leonov). It had been more than 14 years since the launch of Yuri Gagarin and it would be another 20 years before Russians and Americans met up in space once again. On both this and the next rendezvous, a Soyuz would carry the Russian crew into space. *(NASA)*

BELOW An exhibit in the National Air & Space Museum, Washington, DC, the mock-up of Apollo–Soyuz serves as a reminder of the first link in a chain that would lead to the International Space Station and a permanent Russian presence aboard a programme started by the United States. *(David Baker)*

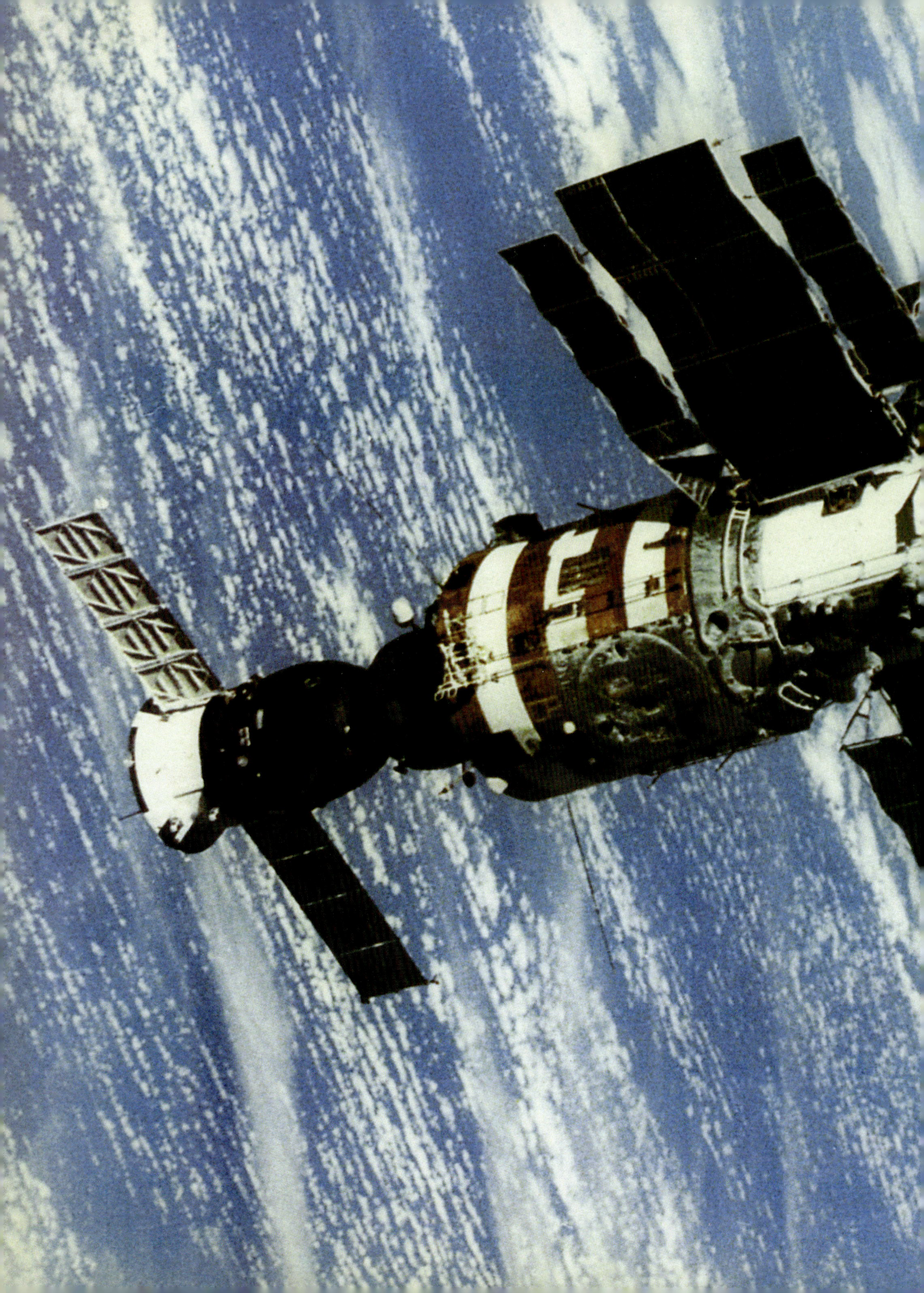

Chapter Six

Soyuz T 1979–86

The last of Mishin's Salyut stations was launched into space on 19 April 1982 and utilised the Soyuz T and TM. Salyut 7 would remain in orbit until intentionally brought back through the atmosphere on 7 February 1991, long after the core for the Mir station had been launched. In that period of more than eight years and nine months, Salyut 7 would receive visits from ten Soyuz spacecraft and Progress cargo tankers.

OPPOSITE Launched in April 1982, Salyut 7 (DOS-6) is visited by Soyuz T-14 in September 1985. Salyut 7 would host ten long-duration missions and received six short-duration crews, some of whom were Intercosmos visitors from non-Soviet countries. *(RKK Energia)*

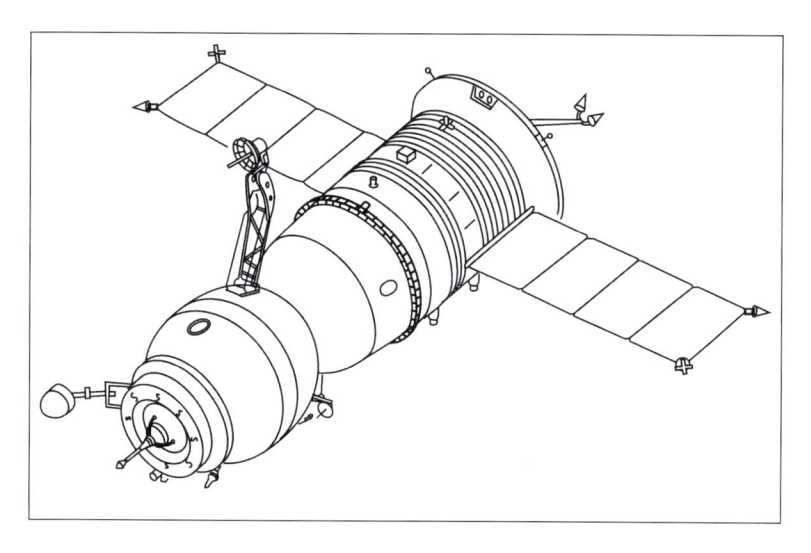

ABOVE First flown with a crew in November 1980, the Soyuz T configuration restored the use of solar arrays, which had been removed after the disastrous flight of Soyuz 11 in 1971. It also reintroduced a three-seat configuration, but this time with suited cosmonauts. *(David Baker)*

BELOW The Salyut 7 configuration with Almaz and TKS module at the front and a Progress transport ship at the aft port. *(RKK Energia)*

The T series adopted solid-state electronics, improvements to the computers on board, including the new BTKSV, and the use of solar array panels that extended the independent life of the spacecraft to 11 days.

With consequent weight savings and modest redesign of the interior layout and systems accommodation, the T series now allowed three cosmonauts in Sokol pressure suits. Redesign of the propulsion system, together with new electronics and control systems, allowed a wider diversity of rendezvous profiles. Also, the orbital module could be left at the station if necessary, although it is believed this was never actually done in practice. Of much greater use, the OM could be jettisoned before the retro-burn, which allowed greater flexibility with propellant reserves at the end of the mission.

Refinements to the descent module included the new controls and displays made possible by the updated electronics and a new Chaika flight control system. The interconnectivity between the ground and the crew in space greatly improved with the Argon 16 computer, and a wide range of small changes, such as the addition of jettisonable window covers to prevent contamination during launch, helped with the overall operability of crew functions.

Incorporating technology developed in the earlier part of this decade, combined with more consistent operational experience with Salyut, the Soyuz T spacecraft derived its suffix from the Russian word for 'transport'. It was no longer just a ferry vehicle but part of an integrated transportation system that now involved the Progress cargo-tanker derivative.

One of the most significant changes came

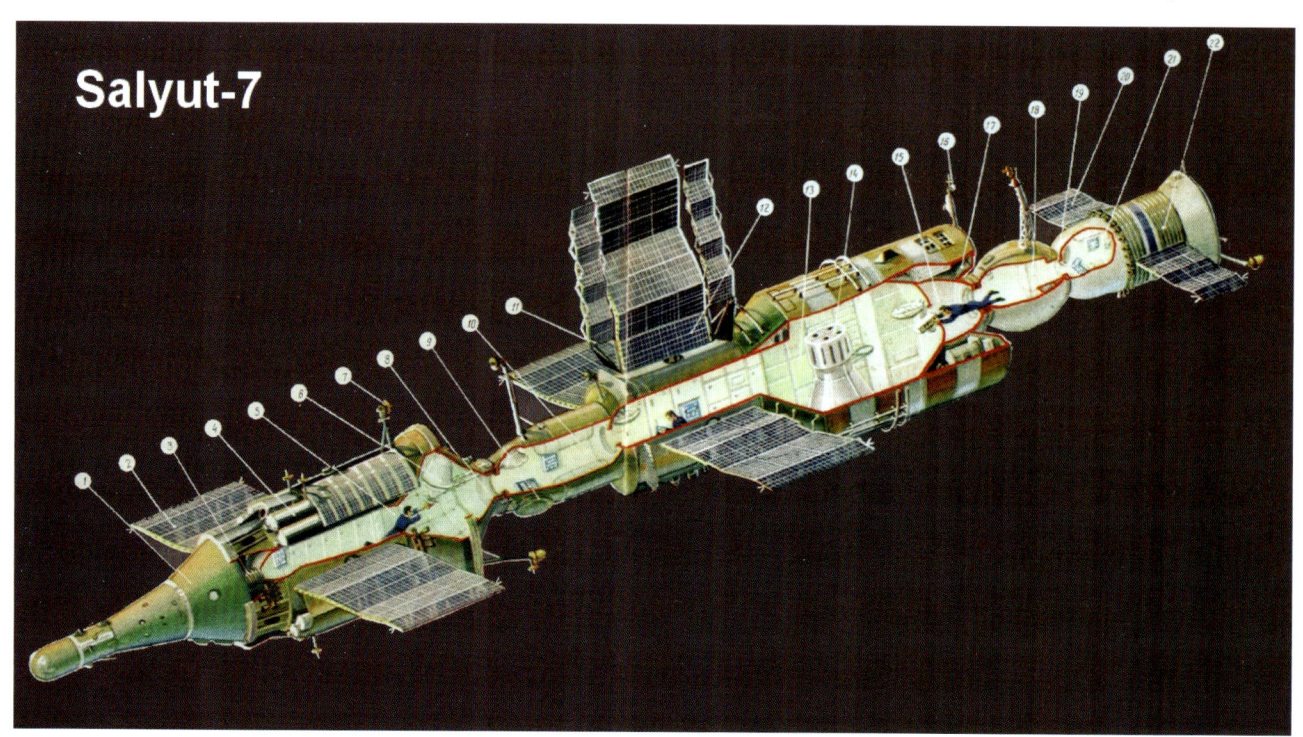

with the reduction in the thrust of the retro-motor through a rearrangement of thrusters systems and orientation engines. Also of high significance was the integration of a common propellant combination for all motor/thrusters systems on board. Instead of hydrogen peroxide for attitude/orientation and nitrogen tetroxide/UDMH for the retro-engine, all motors would use the latter in an integrated propellant flow system so that there could be a budgeted balance between the two.

Three unmanned test flights were flown to test and evaluate the new levels of hardware and software, including Cosmos 670 in August 1974, Cosmos 772 in September 1975 and Cosmos 869 in December 1976. Not all were a complete success, but each helped refine the technology and new operating procedures for two test flights of T series spacecraft. Several more Soyuz flights were flown which added to the growing list of changes to be incorporated into the new spacecraft, not least the changes made for the Apollo–Soyuz Test Project (ASTP), the docking flight with an Apollo spacecraft in 1975.

Two test missions with fully equipped Soyuz T vehicles were flown, as Cosmos 1001 in April 1978 in a none too successful demonstration, and Cosmos 1074 in January 1979 as a 90-day flight curtailed by 30 days due to technical problems. Launched in December 1979 as Soyuz T-1 7K-ST, the first officially acknowledged spacecraft of the new series was flown unmanned on a mission lasting more than three months. Weighing 14,220lb (6,450kg) it was docked to Salyut 6 for 95 days, during which time it remained powered up on Salyut power supplies before two days of independent flight and a normal re-entry.

Launched on 27 November 1980 with Leonid Kizim, Oleg Makarov and Gennady Strekalov on board, Soyuz T-3 was the first to carry three crew members since the fatal mission of Soyuz 11 in 1971, effectively qualifying the new vehicle for its fully defined new role. In all there were 16 launches of the T series, the first (T-2) carrying Yuri Malyshev and Vladimir Aksynonov to the Salyut 6 station for a brief flight lasting 94 hours. The usual practice of leaving the Soyuz spacecraft at the station and returning in a previously docked Soyuz became routine practice and this greatly expanded crew rotation options.

A dramatic start to the flight of Soyuz 10 on 26 September 1983 resulted when the launch vehicle burst into flames 90 seconds prior to the planned lift-off time. The launch control centre commanded ignition of the launch escape rocket, hauling the spacecraft (7K-ST-16L) away from the pad with an acceleration of up to 17g for three seconds, carrying it 2,130ft (650m) into the air.

Vladimir Titov and Gennady Strekalov had no

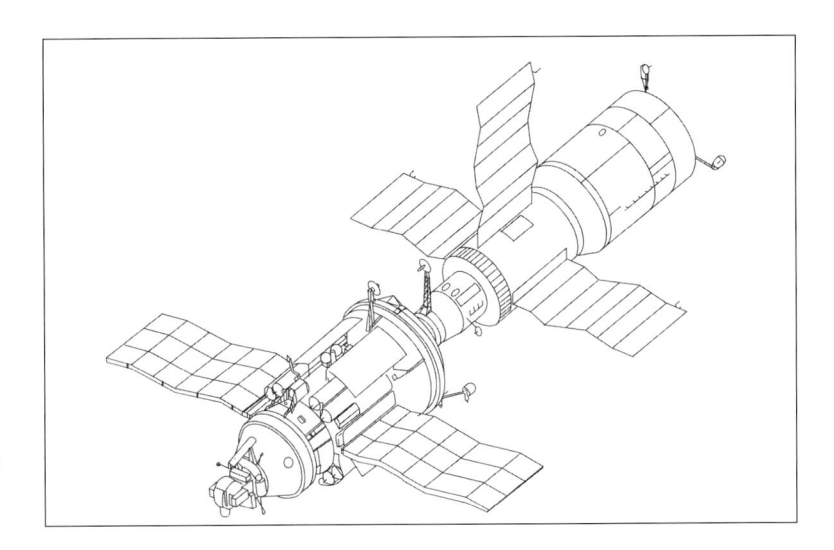

ABOVE In a portent of things to come, the launch and docking of a large module was demonstrated when Cosmos 1686 was docked to Salyut 7 in September 1985, almost doubling the length and mass of the complex. *(Jim Gerrity)*

BELOW A schematic of the Salyut 7 configuration with Almaz TKS return module at the front. *(Charles P. Vick)*

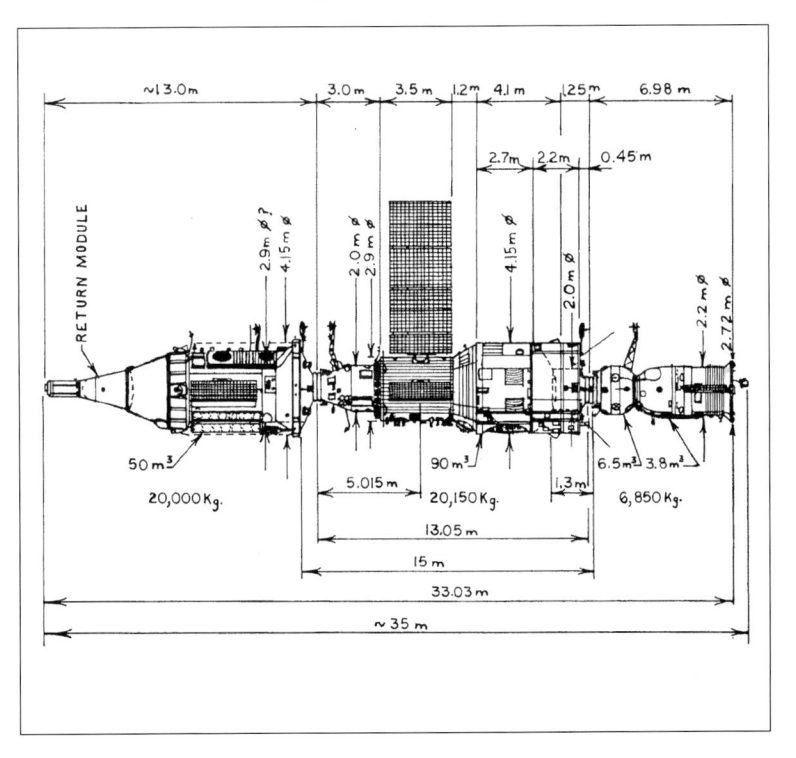

LEFT India's first and only cosmonaut, Rakesh Sharma (left) with Yuri Malyshev and Gennady Strekalov, who accompanied him on the flight of Soyuz T-11 in April 1984. *(Gurbir Singh)*

CENTRE The Soyuz T-10 that brought Rakesh Sharma home had been in space for almost 63 days and is now displayed at the Nehru Planetarium in Delhi. *(David Baker)*

option but to ride out the sequence of events that separated the descent module, deployed the emergency parachute, jettisoned the base heat shield and fired the landing retro-rockets. The crew had only a few minor injuries to show for their 5min 13sec ride. Retroactively designated Soyuz 10-1, the 'real' Soyuz 10 mission began on 8 February 1974, carrying Leonid Kizim, Vladimir Solovyov and Oleg Atkov to Salyut 7 for a mission lasting nearly 63 days.

The last T series Soyuz was launched on 13 March 1986 as a stop-gap measure to maintain flights to the Mir space station in the absence of a new generation TM series spacecraft being available in time. With a weight of 15,100lb (6,850kg), carrying Kizim and Solovyov, Soyuz T-15 remained docked to Mir for almost 51 days before separating and conducting a rendezvous and docking with Salyut 7, to which it remained attached for a further 50 days before translating back to Mir. There it stayed for another 30 days before returning to Earth on 16 July after a total duration of just over 125 days.

Great changes had been made through the T series of Soyuz transport ships and the spacecraft itself had achieved unprecedented levels of reliability. Few could have known that its most effective use would mature with the next series – one which would see a transition not only of Russia from a communist state to a country where free elections were allowed, but also one which would play a great part in uniting world nations in a cooperative endeavour to replace national objectives with an international agenda.

LEFT In 2013, Rakesh Sharma reflects on still being the only citizen of India to fly in space, but he is an enduring inspiration for the young. *(Gurbir Singh)*

Feature D

Progress 7K-TG, 7K-TGM, 7K-TGM1

In the first half of the 1970s, when the future for Russia's human space flight programme seemed to be aligned more with Earth orbit rendezvous and docking flights than Moon missions, several experimental versions of Soyuz were considered. In 1973, while the variant that would be used for the joint ASTP flight with the Americans was being developed, Mishin's team proposed an unmanned version of Soyuz that would carry dry and wet freight to and from Salyut stations. Bearing the designation 7K-TG, such a vehicle was placed in design and in February 1974 a plan was released for detailed engineering and fabrication.

The result was the Progress cargo tanker, a direct derivation of the K-T ferry ship where the descent module was replaced with a structure known as the Otsek Komponentov Dozapravki, or OSD. It would contain propellant tanks and pumping equipment for refuelling the Salyut station together with a mass of 4,080lb (1,850kg). The orbital module would be replaced with a cargo module (CM) for carrying dry freight to the stations, with a structural weight of 5,555lb (2,520kg). Fabrication of the CM would be almost identical to that of the OM in a conventional Soyuz. The instrument module and its propulsion section were largely similar to that on Soyuz K-T, with fourteen 18lb (8kg) thrusters and eight 2.2lb (1kg) thrusters.

Like the manned K-T variant, this first-generation Progress would rely solely on battery power for electrical energy. Because it would be an expendable delivery vehicle, the life of a Progress would be around 33 days of which three days would be required for rendezvous and docking, although some would have an extended capability. Because so much effort had gone into providing the manned Soyuz with automated systems, it was a relatively simple step to transfer that capability to the Progress vehicles, which would operate in much the same way.

The nominal mass of a first-generation Progress was 15,480lb (7,020kg) loaded. Total length was 24.5ft (7.48m) with a maximum diameter of 8.9ft (2.72m). Total load-carrying capacity was a maximum 5,700lb (2,600kg), but usually with up to 2,200lb (1,000kg) for wet cargo (propellant, oxygen, nitrogen and water) and up to 2,850lb (1,300kg) of dry cargo such as could be manipulated through open hatches into the Salyut space station.

The first of 43 first-generation Progress 7K-TG cargo tankers was launched on 20 January 1978 with a full load, of which 2,200lb (1,000kg) was propellant and oxygen. It took several days to unload Progress 1, cargo being attached to a frame structure built into the CM, with smaller items and sundry packages loaded in separately identified boxes. Propellant was carried from the OSD round the CM and into automatic connectors on the Salyut 6 station. Progress 1 undocked on 6 February, withdrew some

BELOW A Progress cargo tanker approaches the International Space Station with wet and dry cargo, replenishing supplies and returning with waste to burn up in the atmosphere. *(NASA)*

155

distance, fired its de-orbit motor and re-entered the atmosphere, where it was destroyed.

The last of the first-generation cargo tankers, Progress 42, was launched on 5 May 1990, returning to destructive re-entry on 27 May. One Progress launch was designated Cosmos 1669, bringing to 43 the total of first-generation types flown. Launched on 19 July 1985, it conducted a routine servicing of Salyut 7 before it carried out a separation and second docking to test automatic equipment, and was de-orbited on 30 August. It is uncertain why this vehicle was given a Cosmos number, as there was nothing unusual about its mission.

The first-generation Progress vehicles were used exclusively with the Salyut 6 and 7 stations, but the planned bigger station – Mir – would require more wet cargo, most of which would be propellant for orbit-keeping activities. This was required to periodically re-boost the station to its appropriate operating altitude; over time Earth's tenuous outer atmosphere would slowly degrade the orbit of the facility and slow it fractionally, which lowered the orbit to a level where, if uncorrected, it would re-enter and burn up. A regular supply of propellant was therefore needed to fire up the propulsion system and raise the orbit back to operating altitude.

Using technologies and improvements developed and applied to the Soyuz T and TM series, a new version of Progress (7K-TGM) was introduced in the mid-1980s and developed as

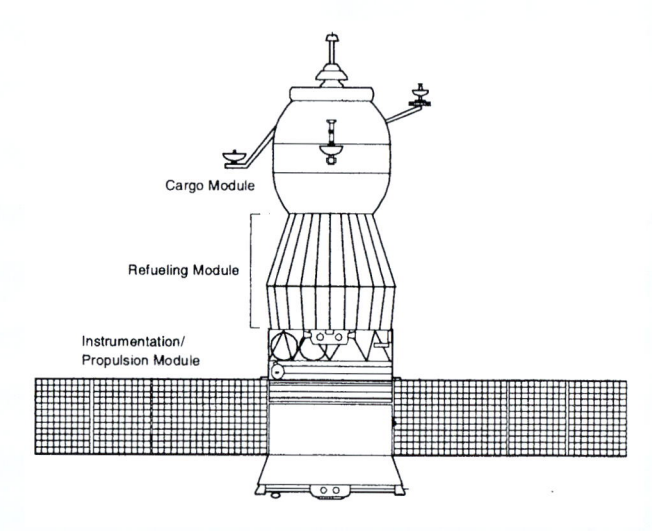

LEFT The general configuration of the Progress vehicle strongly resembles Soyuz, but with the bell-shaped descent module replaced by the pressurised cargo compartment incorporating the refuelling module. *(Energia)*

BELOW The structural arrangement of the orbital module used for dry goods and water tanks as necessary is a structural copy of the equivalent module on Soyuz. *(Energia)*

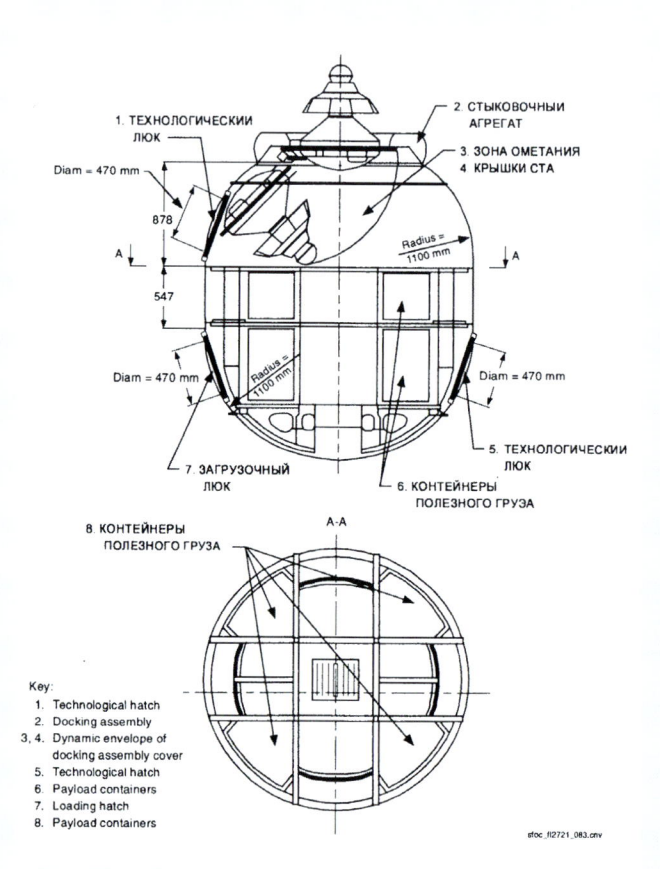

BELOW The instrument/assembly module is relatively unchanged from the Soyuz design and the forward orbital module incorporates the Kurs rendezvous and docking system. *(Energia)*

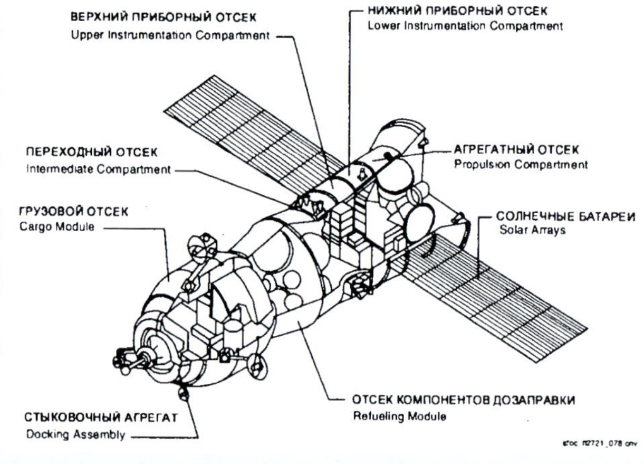

the Progress M, with an extended life of 30 days in automated flight. However, some technologies tried out with Progress M were applied to the manned versions. Unlike Progress, the M series carried solar cell arrays and could accommodate more propellant requirements for Mir. Nominally, Progress M would carry 1,800kg of dry cargo and 2,950lb (1,340kg) of wet cargo, of which 925lb (420kg) was water.

The first of this second generation, Progress M1, was launched on 23 August 1989 to dock two days later with the as yet unmanned Mir station, where it remained until 1 December that year. On 2 February 2003, the 100th Progress launch took place with a visit to the ISS. The last in this second-generation series, Progress M67, was launched to the International Space Station on 24 July 2009 and was de-orbited on 27 September, long after its third-generation successor Progress M1 had started flying supply missions to the ISS.

Originally designed to service Mir 2, which was not ever flown as such but which merged into the International Space Station, further development produced the M1 version in the late 1990s. This had a lower payload capacity, reduced from 5,470lb (2,480kg) to 5,025lb (2,280kg) while retaining the dry cargo of the M series. But within the wet cargo allocation, the number of tanks in the refuelling module was increased from four to eight, raising propellant capacity for transfer from 1,920lb (870kg) to 4,300lb (1,950kg). Changes to the refuelling module in the OSD required the two water tanks to be relocated into the CM.

The first third-generation M1 launch occurred on 1 February 2000 to the Mir station, and the first flight to the International Space Station came with Progress M1 3 on 6 August, when it docked to the ISS, which had yet to begin continuous manned operations. On 24 January 2001, Progress M1 5 was launched to Mir on a station de-orbit mission and remained attached as the entire complex plunged back down through the atmosphere on 23 March that year. The last of 11 M1 flights was launched on 29 January 2004 and de-orbited on 3 June, flights of the basic M series continuing until 2009, as mentioned above.

A subtle designation change came with the launch of a new Progress variant on

ABOVE After docking with the International Space Station, the open hatch of Progress 18 clearly shows the inward-hinging docking probe. Incorporated as an integral part of the forward hatch, the integrated hatch and docking design is common to all Soyuz spacecraft. *(NASA)*

BELOW The Progress M refuelling system for transferring propellant from the tanks stored in the refuelling module to the Russian modules on the International Space Station. *(Energia)*

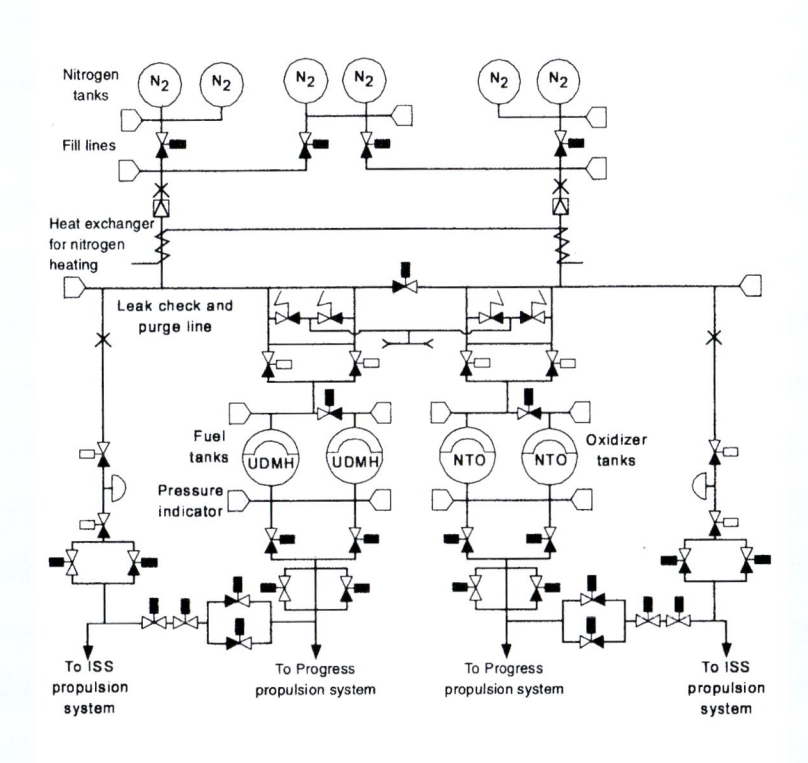

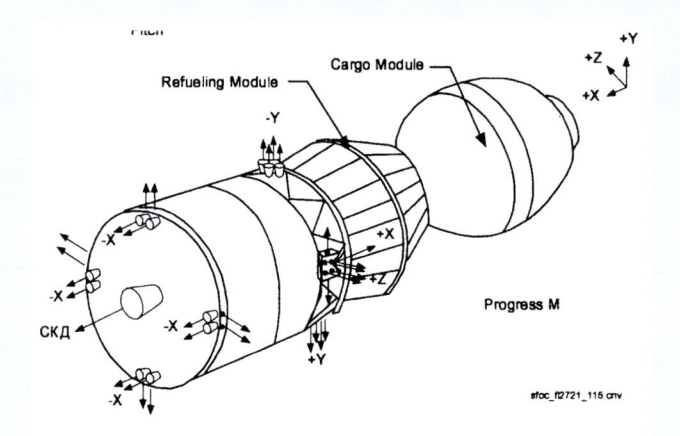

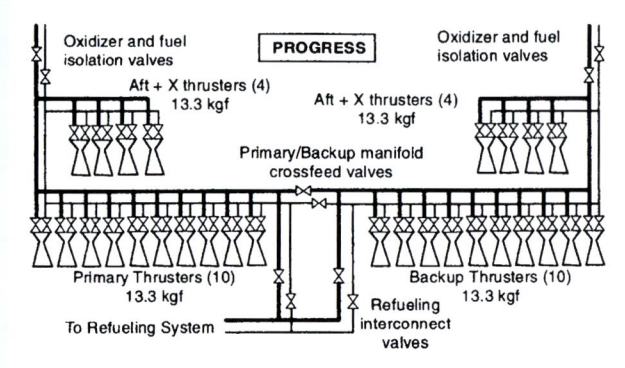

ABOVE Nitrogen pressurising tanks and supply feed to either Progress or the International Space Station is controlled via diverter valves in the refuelling module. *(Energia)*

ABOVE Primary and backup manifold systems on the Progress cargo tanker are shown here for primary and secondary thruster systems. *(Energia)*

BELOW The Progress command and control system is similar to, but not a direct copy of, the equivalent system in Soyuz. This schematic clearly shows the relationship between the Kurs rendezvous and docking system and the functional operation of the control systems for propulsion. *(Energia)*

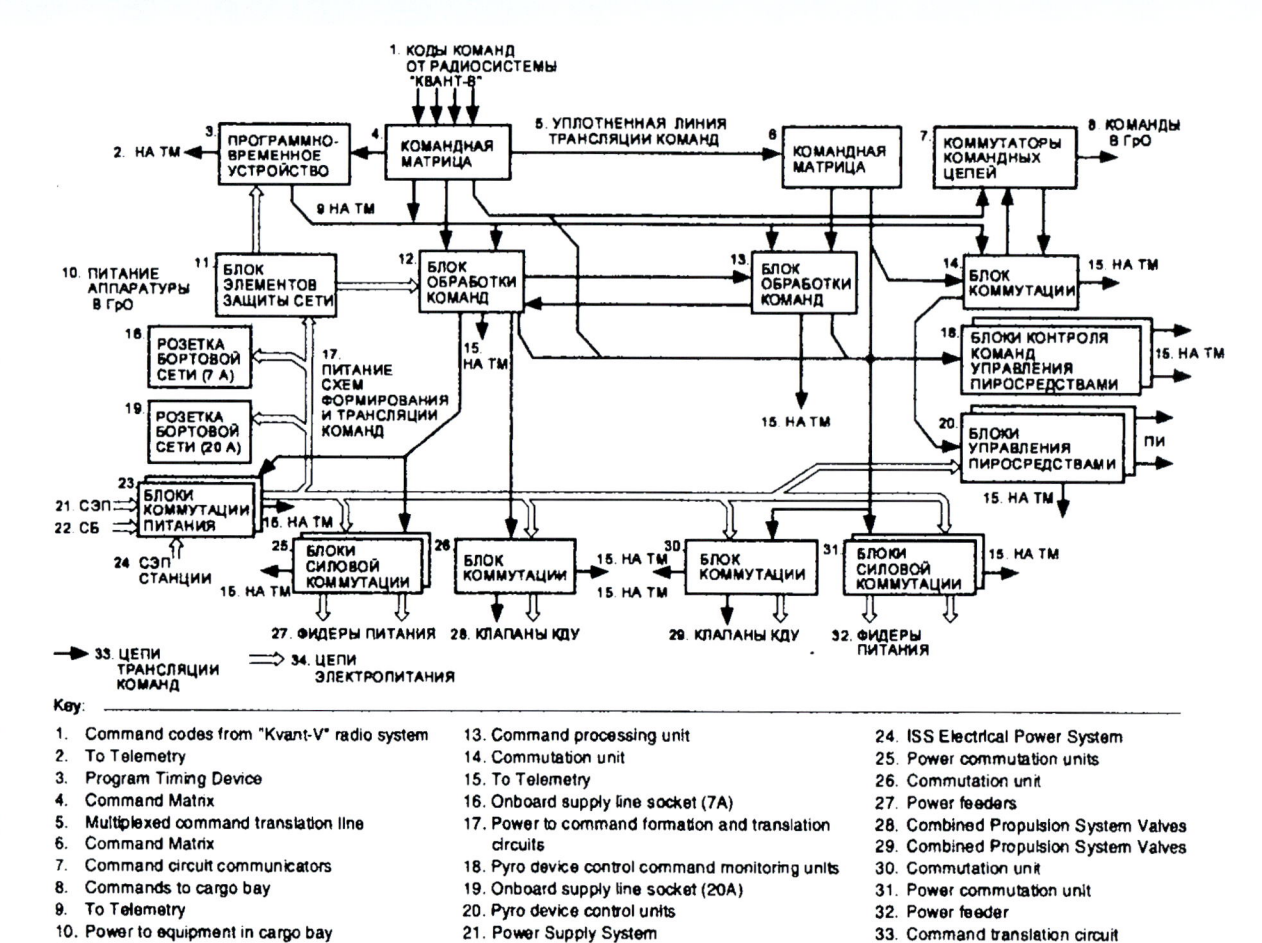

Key:

1. Command codes from "Kvant-V" radio system
2. To Telemetry
3. Program Timing Device
4. Command Matrix
5. Multiplexed command translation line
6. Command Matrix
7. Command circuit communicators
8. Commands to cargo bay
9. To Telemetry
10. Power to equipment in cargo bay
11. Circuit protection element unit
12. Command processing unit
13. Command processing unit
14. Commutation unit
15. To Telemetry
16. Onboard supply line socket (7A)
17. Power to command formation and translation circuits
18. Pyro device control command monitoring units
19. Onboard supply line socket (20A)
20. Pyro device control units
21. Power Supply System
22. Solar Array
23. Power commutation units
24. ISS Electrical Power System
25. Power commutation units
26. Commutation unit
27. Power feeders
28. Combined Propulsion System Valves
29. Combined Propulsion System Valves
30. Commutation unit
31. Power commutation unit
32. Power feeder
33. Command translation circuit
34. Electric power circuits

158
SOYUZ MANUAL

ABOVE A size comparison of NASA's Apollo spacecraft (right), flown with human crews between 1968 and 1975, Russia's Progress cargo tanker (centre) and Europe's Automated Transfer Vehicle. Only the Progress vehicle will remain in use, as the role played by the five ATV cargo supply vehicles will be replaced by US commercial cargo carriers from companies such as SpaceX and Orbital Sciences. *(ESA)*

ABOVE As a demonstration of the flexibility and adaptability of the basic Soyuz design concept, the unmanned Progress M1-4 cargo tanker is seen up close in proximity with the International Space Station. *(NASA)*

BELOW Adapted as a combined delivery truck and docking vehicle, Progress M MIMM02 with the Poisk airlock module being delivered to the International Space Station. The refuelling module and the orbital module are replaced by the airlock assembly which was left permanently docked at the station while the lower section of the Progress delivery vehicle returned to destructive re-entry through the atmosphere. *(NASA)*

26 November 2008, when a modified M1 designated as an M-M type was flown to the ISS. This version carried the then new TsVM-101 digital computer, which was integrated with the MBITS digital telemetry system. To add complexity, these new Progress variants of the third-generation version are designated numerically from that first launch as M-01M, the second as M-02M, etc, and can therefore be referred to as M-M types.

The extreme versatility of the unmanned cargo tanker was utilised for two special Progress M deliveries made to the ISS. The first, M-S01 on 14 September 2001, carried the Pirs airlock module, and the second, M-S02 on 10 November 2009, lifted the Poisk module. In each case both the OSD and CM modules were replaced with the new modules for permanent installation on the Russian section of the ISS.

To the end of June 2014, there have been 23 Progress M-M flights, 11 M1 flights, 69 M flights (including S01 and S02), and 42 of the first-generation type, a grand total of 145 Progress cargo tanker vehicles since 1978. Depending upon how the intervals fall within a calendar year, Progress flights are expected to continue at the rate of four or five flights a year for some considerable time to come.

159

SOYUZ T 1979–86

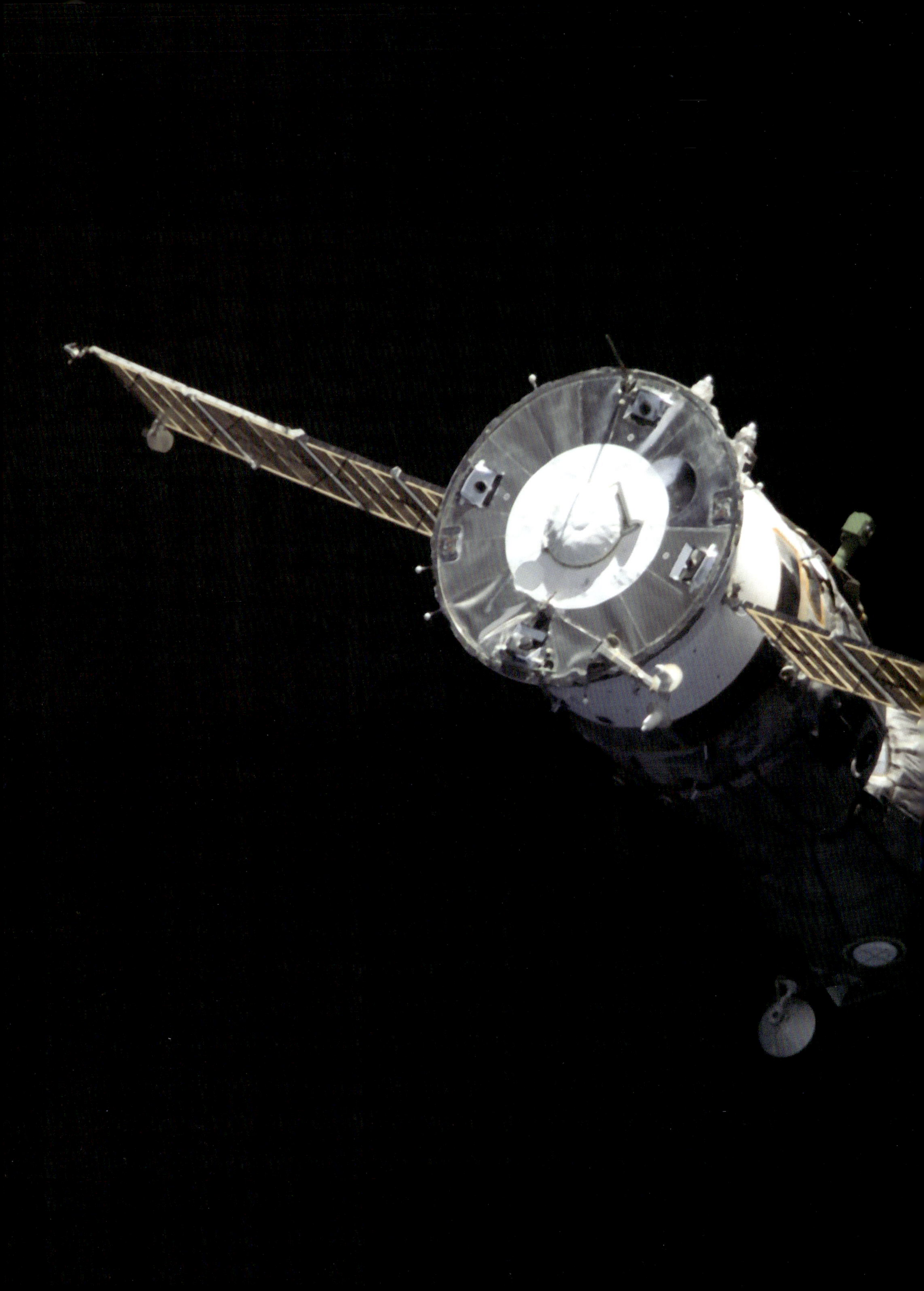

Chapter Seven

Soyuz TM 1986–2002

Further modifications to onboard equipment and spacecraft systems characterised the next Soyuz development, the TM series. Most notable was the shift from the Igla rendezvous and docking system to the Kurs, with the latter's ability to lock on for control computations at a radar acquisition distance of 320 miles (515km).

OPPOSITE Soyuz TM-32 departing Mir showing the after section of the equipment module. The TM-series Soyuz incorporated a forward-facing window in the orbital module and was operational from 1987, making its first manned visit to Mir in February that year at the start of a mission lasting 174 days. *(NASA)*

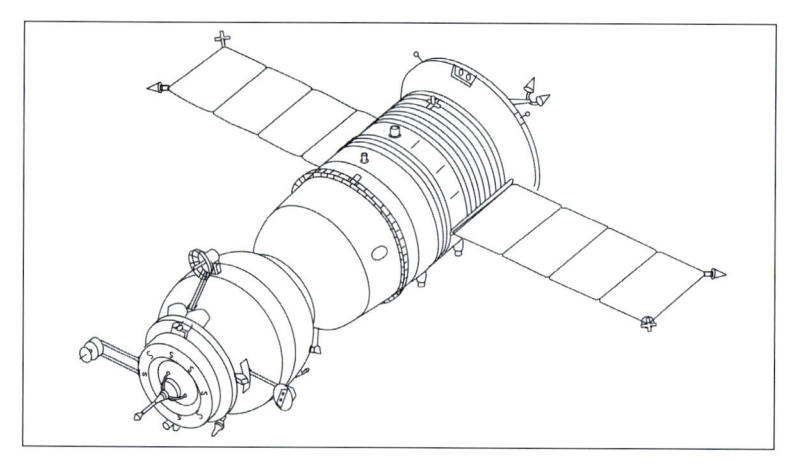

ABOVE The general arrangement of the Soyuz TM remained similar to the T-series, the most notable visual difference being the replacement of the Igla rendezvous and docking system with Kurs, which had a greater acquisition and lock-on range. *(David Baker)*

BELOW Training facilities at Star City have been modified over the years but are still recognisable from the very early days. The circular building at mid-field is the centrifuge facility. *(David Baker)*

For compatibility purposes, the new Mir space stations had the Igla equipment at the aft docking port and the new Kurs system on the forward side port, but when the Kvant module was installed at the rear of the station it too had the Kurs equipment.

A major advantage with the TM was in the addition of a forward-facing window in the orbital module, complementing the somewhat awkward periscope system previously installed in the descent module. This allowed a crew member to be in the OM during the final approach and docking, thus enhancing eyeball manual correction and alignment with the target. Further refinements were made to the Sokol spacesuits and the parachute system was lightened, improved and made more resilient.

The equipment module had several key changes, not least significant an improvement to the materials used for dividing the fuel and the oxidiser in the propellant supply system. There were also some technical design and manufacturing changes to the KTDU-80 retro-propulsion motor, which improved reliability and operation. There were also changes to the electronic control systems to better budget the propellant balance between the retro-motor and the attitude/orientation thrusters.

The Russian space programme was not only evolving through its increasingly updated Soyuz transport ship but also in the sophistication and durability of its space station programme. The appearance of the Mir programme reached a peak during the period when the TM series was introduced, a new level of capability matched not only by the flights to and from space but in the very level of growth with the orbital facility itself.

But it all took time, and with diminishing resources and pressures to develop a new generation station delays were inevitable. The replacement for Salyut 7 was behind schedule and there was a shortfall in the supply of new Soyuz spacecraft. The decision to build a successor to the DOS series of Salyut stations had been signed on 17 February 1976. There were to be four docking ports at the forward end, one at the rear and at the front a series of modules added like four equally spaced spokes on a wheel, with another port at the rear. This port would similarly house a permanent module, itself carrying a docking port for Soyuz or Progress vehicles.

Modules from Chelomei's Almaz programme would be utilised for this new space station and Mir itself would provide sufficient capability for a continuous and sustained habitation over many years. But other work reduced the pace of development, not least a massive effort to build a reusable shuttle called Buran and its Energia launch vehicle, to match NASA's Shuttle. Then, in 1984, came a direct instruction from the Central Committee to have the Mir station ready for launch to celebrate the 27th Communist Party Congress.

The core module for Mir was launched by Proton rocket on 19 February 1986 while operations were still continuing with Salyut 7. As for Soyuz, the unmanned inaugural test of the new series occurred on 21 May 1986 when TM-1 was launched on a nine-day flight to the unoccupied Mir station, returning to a successful landing and clearing the way for manned flight. Operations to the Mir station

RIGHT The Mir training structure simulating the physical layout of the Mir station and, later, the Russian modules of the International Space Station are where much work is done familiarising crew members with the layout of the interior and the location of various items of equipment. *(David Baker)*

began with Soyuz T-15 launched in March 1986 (as recorded on page 154), and this involved dual visits on the same mission between Salyut 7 and the new facility.

Launched on 5 February 1987 with Yuri Romanenko and Aleksandr Laveykin on board, the new Soyuz TM-2 inaugurated its first crew to space, a series which would be the mainstay of successive visits to Mir. On a flight lasting 174 days, the TM-2 was used as a lifeboat in the event the crew had to abandon the Mir station when the unmanned Kvant science module lost lock during rendezvous and threatened to collide with it – an incident only narrowly avoided.

While Intercosmos flights ended in 1988 that was not the last flight to carry non-Russians into space. What may be considered the first in a series of flights that would span the transformation of Russia from a communist state to a democracy began when a group of British companies signed an agreement with the USSR to fly a British astronaut to Mir. With over 13,000 applicants the choice of astronaut would be dependent on physical and psychological tests as well as aptitude,

LEFT The training facility for the descent module has seen a series of modifications over the years as the interior layout and controls configuration for Soyuz have steadily changed. *(David Baker)*

LEFT Incorporating lessons from Salyut stations, Mir was a considerable advance on the early DOS stations that had evolved from their conception in the late 1960s. The Lyappa system at the front docking node allows modules to be swung round 90° from their initial axial docking position to one of the four radial ports, freeing the axial port again for a Soyuz transport ship. *(RKK Energia)*

RIGHT Complete in its fully assembled configuration, Mir hosts the Soyuz TM-24 spacecraft as viewed from the Shuttle *Atlantis*. On all station missions, Soyuz doubled as a lifeboat available to bring the crew home in the event of a malfunction.
(NASA)

BELOW Mir supported a wide range of scientific and engineering experiments which would have resulted in a second-generation Mir had Russia not merged its space station plans with the paper designs of NASA and its international partners in Europe, Japan and Canada.
(NASA)

intelligence and suitability for the demanding and exacting mission the winner would be called upon to endure.

When selected, Helen Sharman was one of the youngest candidates and went through the full training regime in Russia with which cosmonauts had become familiar. She was launched aboard Soyuz TM-12 with Anatoly Artsebarsky and Sergei Krikalev on 18 May 1991. Helen Shaman had functional tasks to conduct in space, including superconductor experiments with the Elektropograph-7K instrument and spoke with British school children in their classrooms live from space. Helen Sharman landed aboard TM-11 on 26 May with Viktor Afanasyev and Musa Mananrov after a flight lasting 7 days 21 hours 13 minutes. As of January 2014, Helen Sharman was the second-youngest woman to fly in space, after Valentina Tereshkova, and is the fifth-youngest person of all time to have ridden a rocket into orbit.

In all, the Russians mounted 30 Soyuz missions to Mir through a time of great transformation, politically and ideologically. With the collapse of the USSR on the last day of 1991, a vacuum appeared, and plans for a Mir-2 were postponed, only to be resurrected on an initiative begun by President Bill Clinton to forge links with the new Russia. Since 1984 the Americans had been planning a space station called Freedom, with the European Space Agency pledging to build a module called Columbus and Japan promising to build their own pressurised module attached to Freedom and called

JEM. The Canadians had agreed to build the robotics for the station.

By the early 1990s very little progress had been made on Freedom and far too little money had been approved by Congress to give it a meaningful schedule for the international partners to work to. On 7 December 1993 an agreement was reached with the existing partners for Russia to join the Freedom programme by adding the Mir-2 core, to which was eventually added a module from Chelomei's Almaz TKS programme. The new facility would be known as the International Space Station (ISS).

TM-31 flew the first Soyuz visit to the incomplete ISS on 31 October 2000, with cosmonauts Yuri Gidzenko and Sergei Krikalev accompanied by NASA astronaut William Shepherd. As Expedition 1 they would begin a continuous human presence on the station that has never ended. For a while the Russians invested in both Mir and the ISS, but on 4 April 2000 Soyuz TM-30 took S. Zalyotin and A. Kaleri on the last visit to Mir on a flight lasting almost 72 days before it was de-orbited on 23 March 2001. Mir had been continuously manned for 3,644 days – almost ten years – a record that was eventually broken by the International Space Station on 23 October 2010.

A new era had dawned, and with Russia's human space flight programme locked in tight to an international programme, soon there would be as many non-Russians as Russians flying aboard Soyuz transport ships. With astronauts from the other ISS partners, astronauts from several European countries as well as Canada, Japan and America were routinely checking in at Star City to follow in the footsteps of Yuri Gagarin and a host of Russian precursors, each one planting a tree to remember the pioneers and never to forget those who had lost their lives in making Soyuz the safe ship it has become.

Freed from rigid communist control, the Russians embraced the ethic of the marketplace as well and, finding willing volunteers, began signing deals for 'private' space tourist flights paid for by wealthy individuals. The first was Denis Tito on TM-32 in April 2001 flying a mission similar to the Intercosmos flights of the 1980s where each visit would last seven or eight days. They would have nothing to do, and their addition to the ISS manifest brought discord between Russia and America as to the suitability for this sort of activity. Mark Shuttleworth followed aboard TM-34 in November 2002, with the historic distinction of flying in the last of the series, returning in TM-33, which had been at the ISS for more than 190 days.

LEFT Britain's first woman astronaut, Helen Sharman flew to Mir in Soyuz TM-12 during May 1991 and returned to Earth in TM-11 after a flight lasting almost eight days, in which she carried out a range of scientific experiments. *(NASA)*

BELOW Soyuz TM-31 is rolled to the pad, the extended tower for the launch escape systems now a dominant feature on the rocket. *(NASA)*

Chapter Eight

Soyuz TMA/ TMA-M from 2002

This latest fully operational version of Soyuz is described in detail in the spacecraft feature on pages 68–97. The primary difference from the previous TM allows a wider percentile limit for cosmonauts and astronauts. This is made possible by redesigned seat liners and changes to the design of the seat struts for bigger crew members. Further improvements were made to the crew displays and panel layouts, with changes to the electronic control systems to improve sequencing between automated events and upgrades to the communications systems.

OPPOSITE Soyuz TMA-20 touches down and remains vertical on landing in Zhezkazgan, Kazakhstan, on 24 May 2011. *(NASA)*

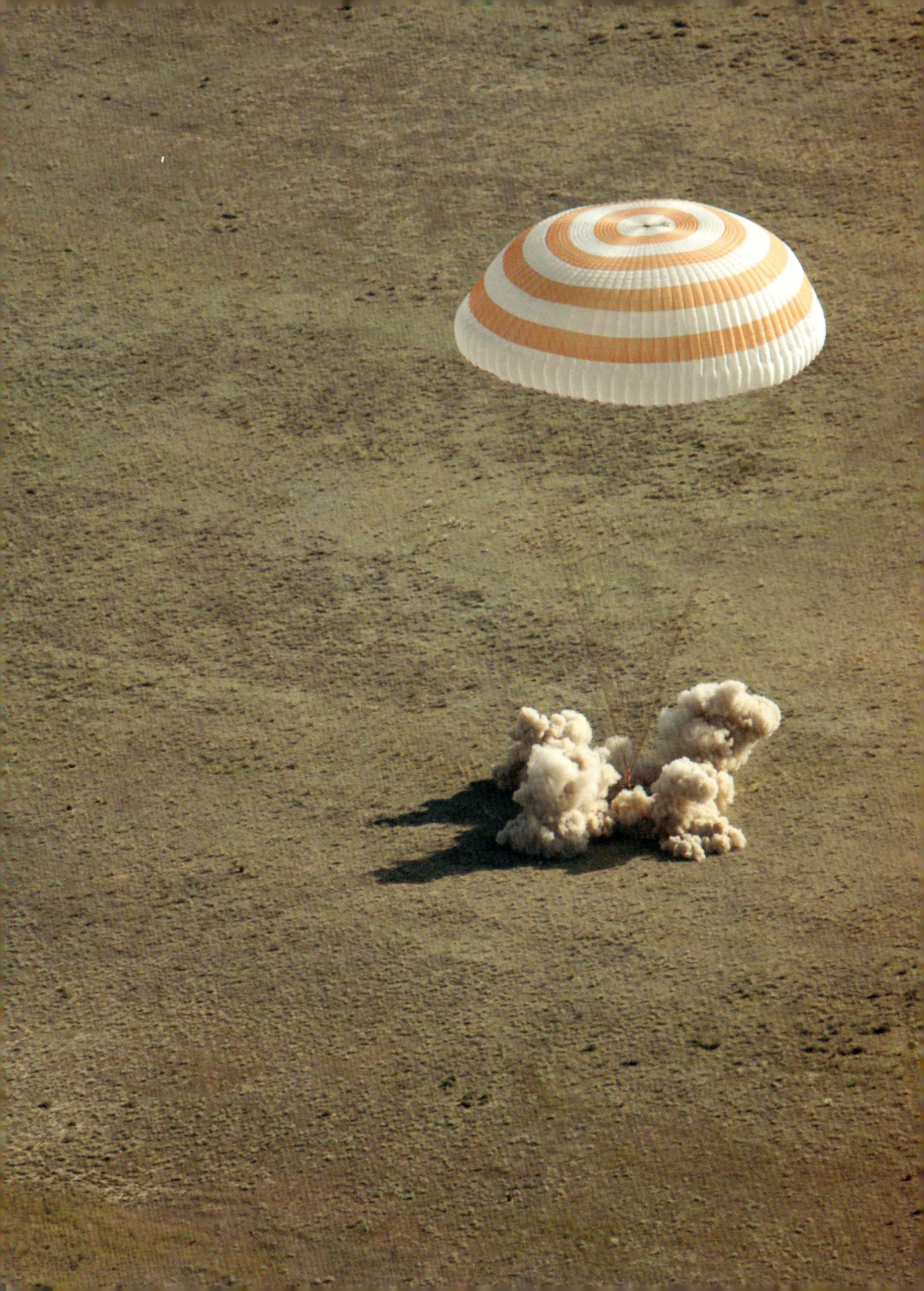

RIGHT Soyuz TMA series spacecraft introduced an improved cockpit interior with provision for larger and smaller cosmonauts, incorporating different seat liners that are always retained by cosmonauts and crew members when they change spacecraft for the return to Earth. *(NASA)*

FAR RIGHT TMA-2 is launched from Baikonur in April 2003 on a flight that will require it to remain docked to the International Space Station on a total space mission of almost 185 days from launch to landing. *(NASA)*

BELOW Soyuz TM-2 returns to Earth in late October 2003. Note the scarred exterior and the parachute container housing. *(NASA)*

The first of 22 TMA flights began with TMA-1 launched on 30 October 2002, and ended with the last flight of this series on 14 November 2011. Each spacecraft remained docked to the ISS for an average 160–190 days, the longest of the series being the 215 days 8 hours 22 minutes of TMA-9 launched on 18 September 2006. It carried up space tourist Anousheh Ansari, born in Iran and now a US citizen, epitomising the complete transformation of Soyuz as purveyor of the socialist ideal, now in part turned taxi for wealthy Western tourists.

But the tourist flights are merely a very useful way of the Russian space programme acquiring currency to help subsidise the expensive training, flight and operations costs involved in running missions to a world-class science laboratory. From the outset ISS Shuttle flights had combined delivery of major structural elements with routine crew changes to keep the station fully occupied.

After the Shuttle *Columbia* was destroyed on re-entering the Earth's atmosphere in February 2003 the Soyuz spacecraft was the only means of delivering people to the ISS and returning them to Earth until Shuttle flights resumed in July 2005. Now that the Shuttle is retired, once again the only way for the international partners to get to and from the station is by Soyuz.

Soyuz TMA-M from 2010

The Soyuz TMA series was developed by RKK Energia, Russia's privatised manufacturer and contractor for space hardware. Its formal name bears testimony to a prestige heritage and a line of great names going right back to the beginning of the Russian space programme. Also known as the S.P. Korolev Rocket and Space Corporation, Energia means 'energy', implying vitalisation and the power to do work, and as such it is a living memorial to its founder more than 55 years ago.

The new Soyuz version incorporated further changes and modifications to the baseline TMA series, with 36 items replaced and a total reduction of 154lb (70kg) in the weight of the spacecraft. A new TsVM-101 computer had been fitted, together with digital avionics equipment and a general reduction in power consumption through improvements to the equipment drawing energy from the solar arrays. Changes have also been made to the materials, including aluminium alloys replacing magnesium alloy on the instrument display.

The first two vehicles, TMA-01M launched on 7 October 2010 and TMA-02M on 7 June 2011, qualified the new series for full operational duty, and with the exception of the last TMA flight in November 2011 all subsequent flights have been with this new vehicle.

For the foreseeable future all manned flights to the International Space Station will begin and end with the Soyuz spacecraft. Operations with the ISS have swung into a cycle where flights are based at intervals of two months and four months, supporting six-month flights with two-month overlaps between Expeditions with their nominal six-person complement.

As of the end of August 2014, the ISS was supporting Expedition 40, to which date there had been 13 flights with the TMA-M version. The most recent launch to that date had been TMA-13M on 28 May 2014, Russia's 122nd manned space flight (the 114th Soyuz flight) starting with Yuri Gagarin's Vostok 1 on 12 April 1961. Whatever the future holds for the development of new or replacement spacecraft, Soyuz is set to continue flying people to the ISS for several more years.

Under current plans there are commercial contenders in the United States that are developing spacecraft for carrying people, and these may supplement Soyuz flights, but not before 2017 at the earliest. Whatever the outcome, Soyuz has a long life ahead and it may soon turn out to have logged the greatest number of space flights from a single design, exceeding even the Shuttle, which made 134 flights to space between April 1981 and July 2011.

ABOVE Soyuz TMA-01M is launched from Baikonur in October 2010 carrying Aleksandr Kaleri and Oleg Skripochka with NASA's Scott Kelly to Expedition 25 on the International Space Station, docking to the Poisk module almost three days later. *(NASA)*

BELOW Expedition crew members Kaleri, Skripochka and Kelly land in snow back on Earth on 16 March 2011, near the town of Arkalyk in Kazakhstan. *(NASA)*

Abbreviations

ACH Attitude control handle.
ARS Atmosphere revitalisation system.
ASTP Apollo–Soyuz Test Project.
CM Cargo module.
COSPAR International Committee for Space Research.
CSM Command and service module.
CTP Thermal control system.
DM Descent module.
DOS Long Duration Orbital Station.
EVA Extra-vehicular activity.
FDAI Flight director attitude indicator.
FDF Flight data file.
GDL Gas Dynamics Laboratory.
GIRD Group for the Study of Reactive Flight.
IAF International Astronautical Foundation IAF.
ICBM Intercontinental ballistic missile.
IM Instrument/assembly module.
ISS International Space Station.
LEPS Launch escape propulsion system.
MOL Manned Orbiting Laboratory.

MOM Ministry of General Machine Building.
NASA National Aeronautics and Space Administration.
OCCMS Onboard complex control and management system.
OCCS Onboard complex control system.
OCS Onboard computer system.
ODCC Onboard digital computer complex.
OK Orbital ship.
OM Orbital module.
OME Orbital manoeuvre engine.
OPS Orbital Piloted Station.
OSD Otsek Komponentov Dozapravki.
PDMI Pressure differential manometric indicator.
RFNA Red-fuming nitric acid.
RNII Jet Propulsion Research Institute.
RTG Radioisotope thermoelectric generator.
SOUD System of orientation and motion control.
TMCH Translational motion control handle.
TMS Telemetry management system.
UDMH Unsymmetrical dimethyl hydrazine.

Index

Afanseyev, Sergei 133
Aleksandrov, S.I. 41
Almaz (Orbital Piloted Station – OPS) 76, 107-108, 116, 119, 133-135, 152-153, 162, 165
101-1 134
Andrianova, Yelena 36
Anti-satellite programmes – see also Soyuz 7K-V1 Zvezda and Soyuz-P 51, 56, 105-107, 120
'Blue' Gemini 106
Bold Orion 106
laser and particle-beam weapons 106
Nike-Zeus 106
SAINT (Satellite Inspection) 105, 120
Hi-Hoe 106
Programme 437 106
Programme 621A 120
Apollo-Soyuz Test Project (ASTP) 122, 136, 138, 140-149, 153, 155
APAS-60 122; APAS-75 122, 140-141, 144; APAS -89 122-124; APAS-95 124
Apollo spacecraft and missions 27, 29, 37, 39-40, 50, 57-58, 60-62, 64, 78, 92, 101-104, 110, 118-120, 124, 126-127, 134, 140-141, 145-147, 149, 152, 159

command and service modules (CSM) 56; CSM-111 144-145, 149
docking module (DM) 142-145, 148
environment 142
fatal fire 100
7 111
8 111-113, 118
9 114, 124
11 115-116
13 141
Associated Press 10
Astronauts 33, 146-147
Aldrin, Edwin 'Buzz' 105, 115, 118, 120, 140
Armstrong, Neil 105, 115, 118, 140
Bean, Alan L. 147
Blackband, William T. 11
Borman, Frank 119
Brand, Vance D. 147, 149
Chaffee, Roger 100
Evans, Ronald E. 147
Glen, John 31.33
Grissom, Virgil 'Gus' 30, 100
Kelly, Scott 169
Lousma, Jack R. 147
Lovell, Jim 119
Shepard, Alan B. 29
Shepherd, William 165

Slayton, Donald K. 'Deke' 129, 147, 149
Stafford, Thomas P. 'Tom' 129, 143, 147, 149
Stott, Nicole 96
Walker, Shannon 96
Wheelock, Doug 96
White II, Edward 48, 100
Automated Transfer Vehicle (ATV) 159
Azov Machine Building Plant 130

Badges 111, 113, 137-138
Baikonur launch facility 26, 57, 101, 112, 114, 134, 168-169
Beria, Lavrenti 7
Blackband, William T. 11
Blagonravov, Professor Anatoli 10, 140
Boardman, John 12
Boeing B-47 106
Brezhnev, President Leonid 36, 40, 63, 101, 118
British Interplanetary Society 37, 139
Brussels World Fair 1958 12
Buran shuttle 122, 162
Busheyev, Konstantin 64

Cape Canaveral 31, 54

Chelomei, Vladimir 52, 55-57, 60, 63-64, 66, 100, 107-108, 119, 133, 162, 165
Chertok, Boris 7, 64
CIA 139
Circumlunar flights 44-45, 48, 50-53, 55-56, 58-61, 63-64, 66-67, 100, 102, 110-112, 115, 117-118, 130
Clinton, President Bill 164
Cold War 63, 138-139
Cosmonauts 23, 25-26, 30-33, 41, 44, 57, 90, 103, 147
Afanasyev, Viktor 164
Aksynonov, Vladimir 153
Andreyev, 147
Artsebarsky, Anatoly 164
Artyukhin, 135
Atkov, Oleg 154
Belyayev, Pavel 41, 44
Beregovoy, Georgy 111
Bondarenko, Valentin 27
Bykovsky, Valeri 23, 34-36, 101
Dobrovlsky, Georgi 131-133
Dzhanibekov, 147
female 31, 34, 164-165
Feoktistov, Konstantin 14, 40
Filipchenko, Anatoly 117, 147-148

Gagarin, Yuri 3, 23, 27-28, 31, 33, 48, 58, 102, 107, 120, 139-140, 149, 165, 169
　death 110
　statue 30
Gidzenko, Yuri 165
Gorbatko, Viktor 117
Ivanchenkov, 147
Kaleri, ALeksandr 165, 169
Katashov, 23
Khrunov, Yevgeny 101, 112-113, 115
Kizim, Leonid 153-154
Klimuk, Petr 134
Kolodin, 131
Komarov, Vladimir 20, 40, 101-102, 111, 116, 124
Kotov, Oleg 127
Krikalev, Sergei 164-165
Kubasov, Valery 117, 147, 149
Laveykin, Aleksandr 163
Lazarev, Vasily 134-135
Lebedev, Valentin 134
Leonov, Alexei 41-42, 44-45, 48, 112, 129, 131, 139, 143, 147, 149
Makarov, Oleg 134-135, 153
Malyshev, Yuri 153-154
Mananrov, Musa 164
Nelyubov, 23
Nikolayev, Andrian 23-24, 32-33, 119
Patseyev, Viktor 131, 133
Ponomaryova, Valentina 34
Popov, L. 137
Popovich, Pavel 23, 32-33, 135
Prunariu, D. 137
Romanenko, Yuri 147, 163
Rukavishnikov, Nikolai 130, 147-148
Sevastyanov, Vitaly 119
Shatalov, Vladimir A. 112-113, 115, 117, 130
Shonin, Georgy 117
Skripochka, Oleg 169
Solovyov, Vladimir 154
Solovyova, Irina 35
Strekalov, Gennady 153-154
Tereshkova, Valentina 34-37, 164
Tito, Denis 165
Titov, Gerhman 23, 30-31, 33, 35, 40
Titov, Vladimir 153
Varlamov, 23
Volkov, Alexander 131-132
Volkov, Vladislav 117
Volynov, Boris V. 112-113
Yegorov, Boris 40
Yeliseyev, Aleksei S. 101, 112-115, 130
Yurchikhin, Fyodor 96
Zalyotin, S. 165
Cosmonauts – non-Russian 137, 151, 163
　Ansari, Anousheh 168
　Gurragcha, Zhugderdemidiyan 137
　Hermeszwski, Miroslaw 137
　Remek, Vladimir 91, 137
　Sharma, Rakesh 154
　Sharman, Helen 3, 164
　Shuttleworth, Mark 165
Cosmos flights – see also Soyuz and DOS
　57 42
　573 134
　613 134
　637 122
　638 148
　670 153
　672 122, 148
　772 153
　869 153
　1001 153
　1074 153

　1669 (Progress) 156
　1686 153
Davidson, Les 10
Design Bureaux
　CKB-29 (Tupolev) 7
　OKB-1 (Korolev) 12-14, 45, 48, 51, 65
　OKB-10 57
　OKB-52 55
　　museum 40
　OKB-586 63
　SKB-385 57
Dogs 25, 36, 44
　Belka 23-24
　Chaika 23-24
　Chemushka 27
　Damka 8, 12
　Kometa 27
　Laika 6, 12
　Lisichka 23-24
　Mushka 23, 26
　Pchelka 23, 26
　Ryzhaya 8, 12
　Shutka 27
　Strelka 23-24
　Ugolek 45
　Veterok 45
　Zvezdochka 27
Dryden, Hugh L. 140

Earth as a Planet exhibition, Kent 1958 12
Eisenhower, President 15
European Space Agency 164

Fatal accidents 23, 26-27, 97, 100, 102, 108, 110-111, 124, 132-134, 140, 145, 148, 168
Faulkerson, Bill 11
Ford, President Gerald R. 148
Freedom space station 164-165
　Columbus module 164
　JEM module 165
Frolov, Yavgeny A. 37

Gagarin Centre near Moscow 92
Gas Dynamics Laboratory (GDL) 6
Gemini programme 36-37, 39-40, 42, 44-45, 48, 51, 58-59, 63, 65, 78, 92, 103, 106, 111, 120, 124, 147
　B 106-107
　VI (6) 32-33, 51, 120, 123; VII (7) 32-33, 51, 120, 123
Glennan, T. Keith 15
Globus IMP 38-39, 79
Glushko, Valentin 6-7, 60-61
Group for the Study of Reactive Flight (GIRD) 6
Halls of Columns, Moscow 24
Huntsville Times, The 28

Institute of Geodesy 10
International Astronautical Foundation (IAF) 139
International Committee for Space Research (COSPAR) 139
International Geophysical Year 1957-58 6, 11, 15, 138
International Polar Year 1882-83 138
International Space Station (ISS) 71, 81, 83, 85, 90, 92-93, 95-97, 121, 125-127, 149, 155, 157-159, 163, 165, 168-169
　Expedition 1 165
　Expedition 24 96; 25 96
　Expedition 25 169
　Expedition 38 169
　Poisk module 126-127, 169

Rassvet module 95
Zvezda module 104

Jet Propulsion Research Institute (RNII) 6-7
Johnson, Vice President Lyndon 29, 140
Johnson Space Centre, Houston 141

Kaluga museum 49
Kamanin, Lt Gen. Nikolai 31
Kantrotwitz, Arthur 106
Kasatkin, A.M. 10
Keldysh, Mstislav 140
Kennedy, President John F. 29, 48, 54, 57, 61-63, 115, 118, 120, 140
Kennedy Space Center 54, 149
Khrushchev, Premier Nikita 6, 9, 17-18, 31-32, 36-37, 40, 48, 55-57, 60-61, 63, 65, 140
Khrushchev, Sergei 56-57, 61
Kissinger, Henry 145
Korolev, Sergei 6 et seq.
　bust 18
　death 44-45, 67
　imprisoned 7-8
Kostikov, Andrei 7
Kosygin, Premier 145-146
Kozlov, Dmitri 57, 104
Kurchatov, Igor 7

Launch Control, Moscow 92
Ledwith, Andrew 10
Lockheed U-2 21, 24
Long Duration Orbital Station (DOS) 119, 130, 133, 162-163
　DOS-2 133; DOS-3 (Cosmos 557) 134-135; DOS 4 135; DOS-6 (Salyut 7) 151 DOS-7K 119
　Lyappa docking system 163
Longest solo flight 34, 36
Low, George M. 143, 146
Lunar LK Lander 47. 55, 59-60, 62, 64-66
Lunar Module 60, 66, 104, 114, 120, 124
Lunar roving vehicles and explorers 61, 116, 130
Luna 16130

Makeyev, Viktor P. 57
Manned Orbiting Laboratory (MOL) 103, 106, 111
Mars 48
Marshall Space Flight Centre 28
Massey, Professor Harrie S. 11
Medals 137, 149
Mercury programme 17, 20, 22-25, 28-31, 48, 58, 92, 129, 147
Mikoyan, A.I. 36
Military-Industrial Commission 64
Ministry of General Machine Building (MOM) 40
Mir space station 76, 80, 85, 92-95, 122-123, 127, 151, 154, 156-157, 161-163, 165
　Kvant module 162-163
Mir-2 space station 122, 157, 164
Mishin, Vasili 45, 67, 108, 112, 119, 130, 133-134, 151, 155
Mission Control, Hosuton 148
Mission Control, Moscow 92-93
Mnatsakanyan, Armen S. 126
Moon landings 29, 41, 44, 48, 51, 58-59, 61, 63-64, 104-105, 114, 118-120, 124, 133, 140, 145-147
　unmanned 116, 130
　L3M programme 133
　N1-L3 proposal 64-65, 100, 103, 108
Moon probes 8, 13-14, 23, 26

NASA 13, 15, 28, 33, 36-37, 50, 54, 58-63, 65, 100, 110-112, 114-115, 118, 120, 124, 133-134, 139-141-143, 149
　visit to Moscow 142, 146
National Air & Space Museum, Washington 149
Nedelin, Marshal Mitrofan 26
Nehru Planetarium, Delhi 154
Newell, Dr Homer E. 11
NII-648 Institute, Moscow 126
Nixon, President Richard 119, 141, 145-146, 149
NKVD secret police 7
North American Rockwell 145
Nudelman, Aleksandr 104-105, 108

Orbits
　changing 51, 58, 120
　Earth 51, 58-59, 64, 66-67, 101-103, 111, 114, 119, 121, 124, 127, 130, 133, 139, 155
　geostationary 51, 83
　lunar 59, 64, 66, 115, 118, 120, 124
　velocity 121
Orion spacecraft 92

Paine, Thomas O. 140
Petrov, Boris 140
Politburo 13
Poloskov, Professor Sergie M. 10-11
Porter, Dr Richard W. 11
Postage stamps 9, 12, 45, 133
Powers, Francis Gary 24

Race to the moon 50, 54, 61, 63, 120
Red Army 7
Rendezvous and docking 48, 51, 58, 65-67, 77, 94-95, 101-102, 108, 110-114, 117, 119-127, 155
　automated 100, 124
　Igla system 117, 124, 126-127, 130-131, 161-162
　Kontakt system 117, 121
　Kurs system 125, 127, 156, 161-162
　lunar orbit 64, 120
　multiple 60
Reshetnev, Mikhail F. 57
Ritchie, Dennis 77
RKK Energia 168-169
　Museum, Moscow 29
Rockets, missiles and launcher vehicles
　Agena 51
　Energia 162
　Redstone 15, 31
　Saturn 28, 105; I 54, 62-63; 1B 1(SA-210) 49; V 54-56, 58, 63, 110, 114, 142
　Soyuz 51-52, 66
　Soyuz U 148
　Titan II 123
　Titan III 103; 3C 111
　A-300 56
　A-600
　ICBMs 7-9, 12
　N1 54-58, 61-64, 114, 116, 139, 145
　R-2A 8, 12
　R-7 6, 8-9, 12, 18, 23, 26, 31, 50-51, 66, 104
　R-16
　R-5663-64
　UR-500K Proton (8K82) 52, 55-56, 63, 66-67, 100, 108, 110-111, 114, 130-131, 133, 135, 162
　V-2 (A-4) 7, 12, 28
　8K72 14, 23-24, 26-27, 54, 56, 105

171
INDEX

8K72K 17-18, 23, 26-27
9K 53-54, 56-57, 60-61, 64-65
11A57 40
11A511 66, 99, 101; 11A511M 104; 11A511U 148
Rocket motors 66
RD-0105 23-24
RD-0108 66
RD-0109 18, 23, 94
RD-0110 66, 94
RD-107-11D511 66
TDU-1 20-21, 26
Royal Society 10
Russian Space Agency 93

Sakharov, Andrei 7
Salyut space stations 80, 126, 130-134, 145, 162-163
1 70, 130-133, 145
2 134-135
3 135
4 135-136
5 136
6 91, 136-137, 153, 155-156
7 (DOS-6) 151-154, 156, 162
Satellites – see also Vostok 139
Explorer 138; Explorer VI 106
Luna 1 14-15; Luna 2 14; Luna 3 14-15; Luna 15 116
Lunik probes 13; Lunik 2 15; Lunik 3 15
Object K – see Vostok
Proton 55
Sputnik 1 (PS-1) 5-6, 8-11, 13, 21, 28, 31, 108, 138, 140;
Sputnik 2 (PS-2) 6, 8, 12;
Sputnik 3 (Object D) 8-9, 12-13
Vanguard 11, 15
Zarya (Dawn) 131
Zenit 18, 31, 104; Zenit-2 31-32; Zenit 4 (Cosmos 59) 42
KH-8 Gambit 57
OD-1 and OD-2 13-14, 18, 31
Sever spacecraft 48, 50-52, 65
Severin, Gay I. 41
Shepherd, Dr Leslie R. 37
Skylab 118, 126, 131, 134-135, 137, 141-142, 145, 147
Smithsonian Astrophysical Observatory 10
Soyuz spacecraft
atmosphere and water supply 71-72
command and control systems 75-88
commanders 93
computers 75-78
communications 72-75, 91
descent module (DM) 79-80, 83, 89, 96, 101-102
first flight 99
flight crew 92
flight data file 89-92
ground crew 92
heat shield 89
instrument module 80
laser rangefinder 90
launch escape propulsion system (LEPS) 93
launching 93-94, 107-108
orbital module 83, 96
propulsion module and system 80, 82-83, 88-89, 95, 100
reference materials 90-91
rendezvous and docking operations 94-95

returning to Earth 95-97
thermal control system 84
toilet 84
Soyuz launches, models and stages
Block C stage 47
Block G stage 47
1 101-103, 108, 111, 116, 124, 140
3 (7K-OK-10) 111, 121
4 102, 112-113, 115, 121, 124
5 (7K-OK-13) 112-113, 115, 121, 124
6 (7K-OK-14) 117
7 (7K-OK-15) 117
8 (7K-OK-16) 117-118
9 (7K-OK-17) 119
10 130-131, 133, 147, 153
10-1 93, 154
11 97, 131-135, 145, 147-148, 152-153
12 131, 133-134, 147
13 134
14 135
16 122, 148
17 135-136
18 135-136
18-1 135
19 122, 144, 149
21 136
23 136
24 136
26 137
28 91, 137
30 137
39 137
11F615 65
3KA 26
7K-OK 47, 65-66, 101, 107-108, 110, 112, 119, 121, 129-130; 7K-OK-1 100; 7K-OK-2 (Cosmos 133) 99-100; 7K-OK-3 (Cosmos 140) 100-101; 7K-OK-4 101; 7K-OK-5 (Cosmos 188) 108, 110, 124-125; 7K-OK-6 (Cosmos 186) 108, 110, 124-125; 7K-OK-7 (Cosmos 213) 110, 124; 7K-OK-8 (Cosmos 212) 110, 124; 7K-OK-9 (Cosmos 238) 111-112; 7K-OK-10 (Soyuz 3) 111; 7K-OK-13 (Soyuz 5) 112; 7K-OK-14 (Soyuz 6) 117; 7K-Ok-15 (Soyuz 7) 117; 7K-OK-16 (Soyuz 8) 117; 7K-OK-17 (Soyuz 9) 119
7K/1L 50-54, 56, 60-61, 64-65
7K-L1 (Zond) 66-67, 101, 117, 119; 7K-L1-2P (Cosmos 146) 100-101; 7K-L1-3P (Cosmos 154) 101; 7K-L1-4 108; 7K-L1-5 110; 7K-L1-6 110; 7K-L1-7 110; 7K-L1-8 110; 7K-L1-9 (Zond 5) 111; 7K-L1-11 115; 7K-L1-12 112; 7K-L1-13 114
7K-L1S 114
7K-L3 64-65, 100, 103, 117, 119
7K-T (T-series) 68, 70, 88, 129-137, 151-154, 156, 162; 7K-T-33 (Cosmos 496) 133; T-1 7K-ST 153; T-3 153; T-10 154; T-11 154; T-14 151; T-15 154, 163
7K-TK 107-108
7K-TM 70, 77, 83, 88, 96, 137, 149, 151, 156, 162; TM-1 162; TM-2 163; TM-11 164-165; TM-12 164-165; TM-16 122; TM-24 164; TM-30 165; TM-31

165; TM-32 161, 165; TM-33 165TM-34 165
7K-TMA 70-71, 88, 120, 169; TMA-1 168; TMA-2 168; TMA-5 134; TMA-7 97; TMA-9 70, 168; TMA-10M 127; TMA-19 71, 95-96; TMA-20 166
7K-VI Zvezda gunship 103-104, 107-108, 111
11K tanker 53-54, 56, 60-61, 64-65
L.4 61
LOK 47 56-57, 59, 63, 67
TAM-M 168-169; TMA-01M 169; TMA-02M 169; TMA-07M 127; TMA-08M 126; TMA-09M 126; TM-11M 169
Perekhvatchik interceptor 56
Progress cargo tanker (7K-TG) 70, 84, 127, 130, 136-137, 151-152, 155-159; 7K-TGM 156
1 155
18 157
42 156
M 157, 159; M-13M 125; M 67 157
M1 157, 159; M1 5 157
M-M 159; M-01M 159
M-S01 159; M-S02 159
Razvedki reconnaissance 57
Soyuz-A 53
Soyuz-FG 94
Soyuz K-T ferry ship 155
Soyuz-P anti-satellite interceptor 56, 107-108
Soyuz-R space station 56, 107-108
Soviet Academy of Sciences 10
Soviet Trade Fair, London 1961 14
Space firsts 37, 48, 101
cosmonauts launched in one vehicle, returned in another 101
docking of two manned vehicles 101
high-altitude abort 135
man-made object to strike the moon 14
manned flight 26, 28, 48, 58, 107, 120, 149, 169
men to sleep in space 30
night landing 131
photographs of far side of the moon 15
satellite 5
spacewalk 45; dual 101
two vehicles docking automatically 108
woman in space 34
Space Shuttle 40, 76, 78, 83, 86, 90, 92, 118, 122-123, 134, 141, 143, 145-146, 149, 162, 168
Atlantis 164
Columbia 149, 168
STS-76 123
Space sickness 30, 33-34
Spacesuits 85-88
helmets 86-87
Orlan 85
Sokol 85-88, 132, 152
Yastreb 112
Spacewalks (EVA) 37, 41-42, 44-45, 85, 101, 121, 132, 142
Stalin, Joseph 7-9
Star City 30, 103, 142, 162, 165
Strategic Arms Limitation Talks (SALT) 146

Tass 44
Tracking ships 83
Tsiolkovsky, Konstantin 48-49, 120
monument 50
Tsybin, Pavel V. 41, 48
Tikhonravov, Mikhail 13
TKS spacecraft 133, 136, 152, 165
Townsend, John 11
Tupolev, Andrei 6-7

United Nations (UN) 11, 139
Committee on the Peaceful Uses of Outer Space 140
US Air Force 106, 111, 120, 139
US Army 15, 106
USSR Economic Achievements Exhibition, Moscow 1960 14
USSR 50th anniversary 67

Vavilov State Optical Institute 13
Venus 28
Volovnik, Lida 14
von Braun, Wernher 8, 15, 22, 28
Voskhod spacecraft 37-45, 47, 50, 64-67, 79, 100, 120, 124
1 (3KV-2) 40, 42
2 39-42, 130
3 44-45; 3KV 37, 39; 3KV-2 (Cosmos 47) 40; 3KV-5 (Cosmos 110) 44-45; 3KV-6 45
4 44
5 44
6 44
Vostok (Object K) spacecraft 13, 17-45, 47-48, 56, 64-66, 112, 120, 124
1 28-29, 31, 169
1K 18, 26-27; 2K 18, 31; 3K 18; 3KA 27, 36; 4K 18
2 30
3 32-33, 45, 119
4 32-33
5 32, 34
6 34, 36
7 51
Block E second stage 18, 23, 94
Korabl-Sputnik 1 (1KP) 23; Korabl-Sputnik 2 24; Korabl-Sputnik 3 26; Korabl-Sputnik 4 (Sputnik 9) 27; Korabl-Sputnik 5 (Sputnik 10) 27
Vykhod spacecraft 37
3KD 37, 39, 41-42; 3KD-4 42

Webb, James E. 61, 63, 140
Weightlessness 23-24, 35, 97, 105, 115, 118-119, 121
Winthrop, Bert 138, 149
Woolley, Richard, Astronomer Royal 10

Yangel, Mikhail K. 26, 63-64
Yelland, Dave 10
Yezhov, Nikolai 7
Young Technology magazine 9

Zond system – see also Soyuz 7K-L1 67, 103, 108, 111, 115, 117
1-3 110
4 110
5 (7K-L1-9) 111, 118
6 112
7 115